FROM MCENERGY TO ECOENERGY

America's Transition to Sustainable Energy

by Dr. Dennis Allen Jacobs and Dr. Karen Anita Branden

WHITMORE PUBLISHING CO.
PITTSBURGH, PENNSYLVANIA 15222

The contents of this work, including, but not limited to, the accuracy of events, people, and places depicted; opinions expressed; permission to use previously published materials included; and any advice given or actions advocated are solely the responsibility of the author, who assumes all liability for said work and indemnifies the publisher against any claims stemming from publication of the work.

All Rights Reserved
Copyright © 2008 by Dr. Dennis Allen Jacobs and
Dr. Karen Anita Branden
No part of this book may be reproduced or transmitted in any form or by any means, electronic or mechanical, including photocopying, recording, or by any information storage and retrieval system without permission in writing from the publisher.

ISBN: 978-0-87426-075-5

Printed in the United States of America

First Printing

For more information or to order additional books, please contact:
Whitmore Publishing Co.
701 Smithfield Street
Third Floor
Pittsburgh, Pennsylvania 15222
U.S.A.
1-866-451-1966
www.whitmorebooks.com

Dedication

To

Joel Jacobs
and
"Oh what a wonderful bird is the Pelican"

Also to those who were kind enough to read
the text and offer some very poignant suggestions,
Especially

Del Corrick
Laurie Desiderato
Mike Armstrong
Gary Jacobs

And thanks to all the friends and family
who showed their support along the way.

Contents

Chapter One: 1
World Energy Overview: A Brief Look, Past to Present
The material in chapter one outlines the content of the text and discusses the present world energy situation. The chapter examines the present state of the fossil fuel industry, primarily petroleum, and briefly discusses the conversion to a sustainable future.

Chapter Two: 6
When Coal Was King: The Beginning of the Hydrocarbon Revolution
Chapter two includes a description of the rise of the coal industry and the technologies responsible for that rise, which produced the Industrial Revolution. It discusses the resulting changes in our cities, cultures, and social systems because of the switch from wood to coal.

Chapter Three: 22
How We Got Here: The Evolution of Petroleitus
Chapter three is an in-depth exploration of the rise of the petroleum industry. It starts at the very beginning with "Colonel" Drake drilling the first well and the boom in the industry from the production of kerosene. It continues with the rise of Standard Oil, looks at the two world wars and how they affected, and were affected, by petroleum. The chapter finishes with the rise and fall and rise again of OPEC and the effect that oil has on modern business, industry, and government.

Chapter Four: 54
The rise of the Utilities: Natural Gas and Electricity
In chapter four, we first look at the development of the natural gas industry, what technology made that development possible, and the government leg-

islation that resulted from and/or helped promote the industry. We then look at the present state of the industry and possible future directions.

Next we examine the scientific and technological discoveries that made the electrical industry possible. We then look at the rise of the electric utilities and the legislation and government policies that accompanied it. Again, we look at the present state of electric utilities and possible future scenarios.

Chapter Five: 82
It's All About Energy
Chapter five contains the most scientific and technological descriptions in the book. In this chapter, we examine the science behind energy industries and look at the various energy sources available to modern civilization in terms of the five steps needed to use those energy sources. The steps are:
1: Extracting the energy in its crudest form.
2: Converting the crude energy into a usable form.
3: Storing the energy until needed.
4: Transporting the energy to where needed.
5: Conversion of the stored energy into a useful application.

Chapter Six: 106
Environmental Consequences of Present and Future Energy Sources
In chapter six, we examine the environmental side-effects of our present energy technologies and sources. The material included in chapter six is based on the latest information that describes the environmental effects on individual and global health. We then examine how renewable sources can avoid some of the environmental problems now facing our civilization.

Chapter Seven: 114
Where Do We Go? How Do We Get There?
In chapter seven, we discuss a possible six-step integrated approach for arriving at a sustainable energy future. The steps are:
1: Raising public awareness to obtain needed support.
2: Implementing conservation measures.
3: Planning for energy storage.
4: Identifying local energy sources.
5: Implementing distributive generation using the local sources.
6: Beginning the process of converting to a hydrogen economy.

Chapter Eight: 124
Social Change and Energy
The material in chapter eight looks at reasons why and how people change from one paradigm to another and how we can apply some of these ideas to raising a consensus that will allow modern society to arrive at a new energy reality.

Appendix A: 134
Various Estimates for the Amount of Petroleum Remaining

Appendix B: 142
The Thermodynamics of Heat Engines: First Law and Efficiencies of Heat Engines

Appendix C: 156
Is It Possible? Can We Go Sustainable? The Amount of Renewable Energy Available

Notes 161

CHAPTER ONE

World Energy Overview
A Brief Look, Past to Present

At this time in our history, we have the opportunity to begin a transition to a sustainable energy future. It is beginning despite our addiction to fossil fuels and the resistance of those who have major investments in our fossil fuel economy. If we can understand and accept the evidence that demonstrates the need for a change and are willing to make the commitment and begin immediately, however, we can minimize the effects that change will have on our society.

The first five chapters of this book will examine the history of the world's energy situation, including the energy sources and technologies available, and the environmental impact of the various technologies. Chapter six lays out an energy scenario that is sustainable into the future, decreases or eliminates environmental problems associated with the fossil fuel technologies, and eliminates our dependence on foreign oil. We will also examine the benefits and possible problems of arriving at such a future and a path to get to that future.

In mythology the Greeks considered one of their greatest gifts from the gods to be Prometheus's offering of fire. That gift to humanity has started a trail of energy development and dependence that has continued for thousands of years. For the first several thousand years, through the 1600s, humans primarily used wood as their energy source. Small amounts of wind and water energy were harnessed for mechanical energy. During the seventeen and eighteen hundreds, when the forests began to shrink, coal became the energy source of choice fueling the Industrial Revolution. Finally, in the late nineteenth and early twentieth centuries, with the development of reliable internal combustion engines, petroleum and natural gas became the fuel to power the twentieth and early twenty-first centuries.

As petroleum became the primary energy source for the modern industrial society, our dependence on it became analogous to an individual addicted to a drug. It is now almost impossible for many of our citizens and leaders to visualize alternative fuels to power our modern civilization. There is a direct relation between adequate oil supplies and our national economy. As supplies dwindle and prices increase, our economy slows down, unemployment increases, earnings decrease, and the average American sees an attack on their modern comforts and lifestyles. An absolute addiction to a resource that has a finite and rapidly diminishing supply can be very dangerous. It is the situation in which the world finds itself at the beginning of the new millennium.

Humans have now consumed over half of the world supplies of petroleum. Depending on which information source we reference, this occurred sometime between the years 1995 and 2005. With the exception of only a slight decrease in consumption during the oil embargo of the seventies and again in the early eighties, the world population has increased its oil consumption every year since the early part of the twentieth century. The increase has been almost tenfold since 1950. It is estimated that within the next five years, the world demand for oil will exceed the world's ability to produce oil. Also, the North American demand for natural gas will exceed our supply of NG within the next ten years. In other words, demand will outstrip production. This does not mean we will run out of oil or natural gas in the next five to ten years, only that it will become much more expensive and harder to find. In the very near future, there simply will not be enough petroleum or natural gas to supply the ravishing appetite of our addiction.

Some have argued there are large sources of oil in yet undiscovered or untapped reservoirs despite evidence to the contrary. Most of the world has been tapped for the easily available oil. Much of what remains is available only in extremely difficult areas to access or in environmentally sensitive areas such as the Alaskan wilderness and coastal California. At current rate of usage, the oil available in the Alaskan wilderness would supply the American petroleum appetite for a maximum of eighteen months. The coastal California supply would last for less than three months.

When discussing oil reserves with the general public, there are one of four scenarios that are invariably brought up. The first is that there are billions of barrels of oil in the Canadian Tar Sands that are at our disposal. The second is that there are billions of barrels of oil in oil shale in the western states that, when the price is right, will be at our disposal. The third is that there are billions of barrels still remaining in the southern oil wells that, when the price is right, will be available for use. The fourth is that market forces will take care of the problem. When oil becomes too expensive, other energy sources will be developed that will save our economy. The first three of these will be discussed later in chapter five. Here we will look briefly at market forces.

To this time, our population has been relying on market forces to solve our energy problems and up to this time, they certainly have not solved any environmental or energy problems. Market forces refer to the economic law of supply and demand. If a product is in demand, the price will remain high until a surplus of the product is available and the price goes down. The price continues to decrease until a balance between supply and demand is achieved. If the supply continues to exceed the demand, the cost will shrink to almost nothing. If the demand continues to exceed the supply, the cost will continue to rise.

When demand is high, various mechanisms are typically employed to increase or continue production, including better manufacturing techniques, increased raw materials, more efficient design, or new technologies. Underlying all of these mechanisms, there is the assumption of sufficient raw materials to continue or increase production. The following chapters describe how each of these scenarios have been repeated over and over in the energy industries with no real foresight or long-term planning. Because market forces are *reactive*, not *proactive*, they typically make changes over long periods of time. At times market forces will react to problems with new technologies or energy sources as we see with the development of the steam engine to pump water from coalmines or using gasoline-powered transportation instead of horse and carriage. These are all reactive forces, however, and they have taken from fifty to one hundred years to evolve.

At this time in our history, we simply cannot wait for market forces to solve our problems. It would be disastrous for our economy and environment. We have already seen these effects in the 1970s during the OPEC oil embargo. A decline of only 5 percent in our oil supplies sent our unsuspecting economy into a tailspin, produced waiting lines at the pump, and produced large-scale unemployment. According to most in-depth forecasts, if we rely on market forces to prepare us for the future shortage, our economy and environment will be in very deep trouble.

What's in Store for the Future?

No one really knows for sure what is going to happen in the future. We do know some basic facts and from these facts, we know some things are going to change. First, we know we are running out of petroleum with very few, if any, world reserves left to tap. Second, we know our transportation system is based on oil. Third, we know our agricultural system is based on oil. Fourth, we know most of our industrial and commercial well-being is dependent on oil. Fifth, we know there is an almost one-to-one relation between our economic health and a continuous oil supply. Sixth, we know the fossil fuel economy is leading to armed conflict and large-scale environmental problems, including pollution, health problems, and global climate change, which will have disastrous effects on Earth's ecosystems. We know, therefore, that the above situation is not sustainable and cannot continue. We, as a nation and a world, cannot continue on the same path we have been on for the past one hundred years.

Something Has to Change.

Does this mean we will see radical changes in our civilization over the next few years? There certainly will be some changes, but if we are proactive and begin to plan now, the negative effects of these changes on our economic and social fabric can be minimized and we can begin to reap the positive benefits. We will we see dramatic increases in fuel prices over the next few years. This increase in fuel prices will force some of the increased efficiencies that are needed. As a world and as a country, however, if we are to avoid major economic and social upheavals, we must do more than just wait for market forces to bring about changes in our energy sources and usage. There will most likely be changes in the power structure of the energy suppliers. We most certainly will not be able to maintain the present American gas-guzzling transportation system. We will need to be more efficient and less wasteful in our space and water heating. Industry and agriculture will need to become more efficient.

As we will see in the following chapters, it will be difficult to look toward our national political leaders for help in this transition. There is simply too much money and political influence on a national level by the fossil fuel industries. We must begin the transition into a sustainable economy at the local level.

Let us briefly look at the positive and negative effects of taking the path to a sustainable future in contrast to the effects of taking no action. If we plan for a sustainable future and implement that plan, it will create more jobs in the energy sector for Americans and fewer of our energy dollars sent overseas, less dependence on foreign oil, more security for our economy, better balance of payments in our foreign trade, tremendous environmental benefits, and a better world for our children and their children. The downside will be inconveniences, from changes in our massive energy-guzzling way of life because of conservation measures that must be implemented. If we do nothing, we run the risk of more military clashes over oil, a continued exporting of our energy dollars, even more massive imbalance in our foreign trade, major damage to our economy as oil supplies decrease, continued environmental damages as a result of fossil fuel burning, and a much less secure world for our children. We simply must take the path to a sustainable future.

In fact, there are positive activities already occurring that will help usher in the next phase of our energy evolution. Wind power now supplies the world with ten times the amount of energy it did ten years ago and is the fastest-growing source of energy in the world. By the end of 2006, the United States had installed 9,149 megawatts of wind generators, which supplied 24.8 billion kwhrs of energy. The cost of wind-generated electricity is now almost equal to coal and petroleum generated electricity, and if environmental costs were included, wind would be far cheaper than fossil fuels.

The use of solar power, both photo voltaic for generating electricity and directly for water and space heating, is also rising dramatically. As the use of photo voltaic systems has increased, the cost per watt has dropped to about one-half of the price it was ten years ago. This process should continue into the foreseeable future.

What's the Problem?

Why then, with wind-and solar-generated electricity growing rapidly, can we not just wait until market forces mandate the transition to the solar economy? The next few chapters will show that finding the energy is not the problem. There is enough wind in the Dakotas and Minnesota to supply all of our electrical needs in the United States. If we combine that with the other renewable sources available across the country, we have plenty of energy to continue our modern civilization. As we will see in chapters five and seven, the problem we are facing as a nation and a world is not that of finding the energy but of *storing and transporting the energy.* It is in this storage and transportation infrastructure that we must look for local solutions. We must quickly begin the planning at the local level to find solutions to these problems if we are to continue as an industrialized nation.

The following chapters will show the history of how and why we got into this situation, why petroleum has been the fuel of choice for the past one hundred years, and a look at alternative fuels and some of the problems with each, and they will finally discuss a plan for the evolution into a new type of sustainable energy economy.

Chapter Two

When Coal Was King
The Beginning of the Hydrocarbon Revolution

Author's Note: Chapter two briefly discusses the beginning of modern civilization and the part coal played in that beginning. If the reader is interested in further study in this area, one of the better non technical books covering all aspects of coal is *Coal – A Human History*, written by Barbara Freese.

At this time in our history, perhaps no energy source is more controversial than coal. It is the darling of many who feel it must become the primary fuel for generating electrical power as our petroleum supplies dwindle. Others feel its negative environmental impacts far outweigh any advantages it has as a fuel. We all have mental pictures of coal powering the Industrial Revolution, where women and children slave over machinery while others work in the mines to supply the black rock that powers those same machines. We have visions of whole cities blacked out in the middle of the day from soot produced by coal fires and young children with bowed legs from rickets brought on by the lack of sunlight obscured by that deadly smoke.

Modern coal enthusiasts claim that vision belongs in the past and with modern designs and combustion technology, coal can be just as clean a fuel source as natural gas. In this chapter, we will examine the history of worldwide coal use and briefly look at the modern combustion technologies. We are now at a cusp, a critical juncture that will determine the energy supplies, both in the U.S.A. and worldwide, for the next century. If we are to choose the correct paths, we must determine if the claims made by the coal enthusiasts or the environmentalists are valid. Then we must decide the type of future we want for ourselves and our children based on that information.

Coal is certainly the most abundant of the fossil fuels and even though it has been in use for the longest time, it still appears to have the largest reserves. Depending on the source one checks and how fast it is used, it appears those reserves should last anywhere from fifty to two hundred years.

Coal also has the most varied history of all of the fossil fuels. Like the other fossil fuels, it is stored solar energy. Most of the coal was formed around 360 to 290 million years ago during the Carboniferous Geologic, or coal-forming period, primarily from very large trees. The harder coal, anthracite and bituminous, was formed during the earlier periods with the softer lignite being formed later and peat still later.

Although there is some evidence that early cave dwellers used coal as an energy source, the first recorded use is by the Romans. They found it attractive and easy to carve as well as combustible and used it both in making jewelry and as a fuel. England fell into the dark ages in the fifth century once the Romans left. Very little was heard of coal until the thirteenth century. During the 1200s, English blacksmiths began using coal to fire their limeburners, which left a pungent, foul odor that offended a large part of the population, primarily bishops, barons, knights, and other nobles. In 1306 King Edward I banned its use, but the ban was largely ignored for the simple reason that wood fuel was becoming difficult to find.

Here we see the beginnings of the social, Industrial Revolution that was to change the world. Previous to coal, the primary fuel to power the world's energy demands was human, horse, oxen, and wood. For mechanical work, human, horse, oxen, and later water wheels sufficed, but for heating, wood was the only choice. In the Northern Hemisphere, the size of cities was largely determined by the availability of wood. As a city developed, it used the surrounding forests for heat energy. As the forests diminished, new cities and towns were built closer to the forests that would supply their energy, thereby limiting the size of towns. Once a seemingly endless supply of heat energy/coal was found in a given area, there was no energy limit to the size cities could become. Over the next three hundred years, through various accidents of history, several cities in England would arrive at this seemingly endless supply of heat energy.

Coal was found, and mining began along the River Tyne, near Newcastle, in the northern part of England around A.D. 1200. Although there were other coal sites in England during that time, the Newcastle site became the most important. First, it was easy to mine; second, the coal could be moved downhill to the river with little effort; third, and most importantly, the coal could then be distributed over much of eastern England, primarily London, on barges floating the river Tyne.

Fig. 2.1. Newcastle is located in North Umberland near the Scottish Border. Coal was shipped by barge throughout much of eastern England.

From the 1200s through the late 1500s, the coal area around Newcastle was controlled by the Roman Catholic Church, who did not want to invest in a major mining operation that would be needed to fully develop the mines in that area. Henry VIII solved this problem when he decided to divorce Catherine of Aragon, which the pope would not allow. Henry broke with Rome, confiscated the church's property in England, and sold it off to the growing class of merchants. The merchants were more than willing to expand the mines and supply the cities with their much-needed energy.

Most of Europe, including England, had by this time entered into a little ice age that would last until the late 1700s. This, along with the rising population, was causing rapid decimation of the forests. Queen Elizabeth, Henry's daughter, sent out a number of commissions to study the wood shortage. They all found the forests were being destroyed, which was not only a problem in the energy area but also in the construction business, iron manufacturing, and in the beer brewing industry. It was becoming very difficult for Londoners to heat their homes using wood. If the heat energy supply problem had continued, the population of the larger English cities would have been decimated through disease. Coal, however, would come to the rescue and by 1600, the population of London had reached two hundred thousand and by 1750, it would be the largest city in Europe. This progress was not without its problems. By the mid-1600s, the sun could barely penetrate the thick coal smoke covering London, and travelers could smell the smoke miles before they reached the city. Londoners' clothes were constantly smoke- and soot-filled and so they began carrying black umbrellas. Bees and flowers began dying off, and London gardening almost came to a standstill except for the few plants hardy enough to withstand the smoke.

We also see at this time the beginning of the confusion in the effects that coal smoke has on human health. There was little doubt that the soot and smoke was filthy and smell was nauseating, but beyond that very little was known. Chimneys were now being installed in most houses to keep the dense smoke out of the home. In the early and mid-1600s, several books and essays were written on the topic, some in favor of the use of coal and some describing the problems with air quality. Perhaps the best know and most vivid is a book called *Fumifugium* written in 1663 by John Evelyn. Evelyn paints the picture of a dismal London with the sun barely able to penetrate the thick cloud of smoke hovering over the city. In these writings, we find quite a lot of experiential evidence for the effects of smoke; however, very little scientific evidence was gathered concerning the negative effect of smoke on human health.

Here we also see the beginnings of accumulation of statistical records on causes of death. A Londoner by the name of John Graunt began to compile information concerning reasons for death and would become known as the father of statistics and demography. Since the 1500s, women called searchers would inspect the corpses of the London deceased to determine the cause of death. These records were maintained to inform the wealthy when the plague was prevalent so they could leave London. Graunt determined the data would be more open to analysis if displayed in tabular form. He published his work in 1662. Although the potential for analysis was there, it would be several centuries before scientific determination of the causes of death would allow for valid interpretation of the data. Causes listed such as grief, itch, piles, planet, rising of the lights, and mother were a bit vague. The tables do, however, show an extremely large proportion of child and other deaths due to lung disease.

Even if the Londoners were able to better determine what the burning of coal was doing to their health and really cared, they probably would have stuck with coal anyway, for the simple reason there was no alternative at the time. There was not enough wood to heat their houses, so it was simply a choice between slow death by pollution or a quicker death by freezing.

The Revolution Continues

Although the transportation and burning of coal allowed for the growth of population in cities, it was another use that would have the most lasting effects on human civilization. As mention previously, coal began to replace wood as a form of heat energy in the thirteenth century. For mechanical energy, however, there was still only the human, horse, oxen, and water wheel, which were very limiting forms of mechanical work. In the late seventeenth and early eighteenth centuries we see the beginning of a revolution that would change the very way humans interact with their world. We would forever be changed.

James Watt is usually given credit for the discovery of the steam engine. In reality there is no one person who can claim to be the parent of steam power. As with most technological innovations, there is a gradual evolution of improvements that led to the use of steam to power the Industrial Revolution. In this section, we will follow the trail of discovery from the first early attempts to the modern steam turbine and see their effects on civilization.

The first use of heat to expand gases to produce mechanical work is credited to Hero at the Great Library in Alexandria around 150 B.C. He developed several variations of a form of steam power. In most variations, steam was produced in a boiler and fed through tubing to two counter rotating orifices. When the water turned to steam, it would expand and escape through the orifices, producing a rotational motion of the tube and orifices (mechanical work). Hero wrote several papers that compiled the existing knowledge on steam, air pressure, and vacuums, and they were later gathered into a treatise on pneumatics. The writings demonstrate a very sophisticated understanding of gas pressures, and some of the ideas were applied to designing flow-control devices and using gas pressure to open and close doors. As with most of the knowledge developed at the Alexandrian Library, the work on pneumatics was lost until the Renaissance. Many of the ideas that led to the rebirth of Europe were the rediscovery of the work developed at the Great Library.

Eighteen hundred years later, in the early 1600s, we see Galileo arguing that a vacuum can exert a force that can be used to lift water out of well. This is a common misconception held even today by many who claim a suction force lifts a fluid through a straw. Upon Galileo's death, his student, Evangelista Torricelli, explained correctly that air exerts a pressure and by removing the air in one area, a low-pressure area (partial vacuum) is produced. The outside air pressure will then exert a force to try to fill that low-pressure area. Although crude vacuum pumps had been around since the Greeks, this correct explanation led to the further development of vacuum pumps to move water.

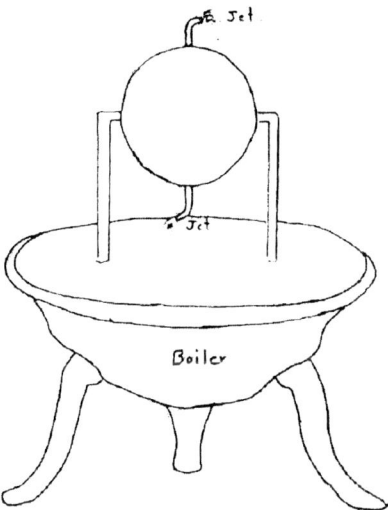

Fig. 2.2. Hero's steam engine. Water is placed in the boiler below the sphere and heated. Steam is generated and passes through the tubes from the boiler to the sphere. As the steam exits the bent tubes inserted into the sphere, it will produce rotation.

By the late 1600s, the demand for coal was creating major problems in most of the coalmines. As miners dug deeper following the coal seams, they began to encounter deadly gases such as carbon dioxide, carbon monoxide, and methane. These gases would kill miners through oxygen deprivation, poison, or explosion. The mining accidents caused by the gases became so prevalent that mine owners asked the newspapers not to report accidents and fatalities because, as the newspapers reported, it would "have very little good tendency, we drop the further mentioning of it." Coal mining continued with the high fatality rates because there were large profits to be made.

Another problem faced by miners and owners could, however, halt the mining operations completely, that of water filling the mines. As the mines went deeper, rainwater and seeping groundwater began filling the shafts, preventing the removal of coal. Because of loss of profits to the owners and the absolute need for coal in the cities, there was an intense search for solutions to the water problem. In Newcastle most of the mines were located on hillsides, and initially drainage tunnels could be dug to drain the water from the shafts. Once the shafts were dug below the adjacent valleys, drainage tunnels were no longer a solution and other methods had to be found. Buckets were used and either hauled up singly on the backs of miners or through a windlass that hauled a series of buckets up on a common rope. By the early 1600s, the granddaddy of the vacuum pump began appearing in some mines. It was usually powered by horses, humans, windmills, and for some mines with more fortunate locations, waterwheels would pump the water. The race was on to find a machine that could power the pumps to remove water from the coalmines.

Europe, primarily Northern Europe and England, was entering the Age of Enlightenment. The Renaissance had brought Europe out of the dark ages and with Newton's publication of *Principia,* a new age in understanding of and working with nature was unfolding. A firm belief in the principle that understanding the laws of nature could bring about machines for accomplishing any purpose was developing. Some of the greatest minds in England were attempting to apply science and technology to solving the problem of removing water from the mine shafts.

We see the first steps taken in the latter 1600s by Christian Huyghens, a Dutch scientist. In designing the first piston-driven engine, he demonstrated that exploding gunpowder (expanding gases) could move a piston through a cylinder, thereby doing mechanical work. Huyghens probably got this idea from the rifle or cannon, where exploding gunpowder moves a projectile through a cylinder.

One of his associates in the design of the gunpowder engine, Denis Papin, translated the concept of a piston moving through a cylinder into the design of the first steam engine. The Papin steam engine was simply a metal tube (cylinder) closed at one end with a piston inside. Water was enclosed between the piston and the closed end with the other side of the piston open to the atmosphere. When the cylinder was heated, the water would turn to steam and the expanding gas would move the piston upward along the cylinder. Cold water was sprayed on the tube, condensing the steam back to water, decreasing the pressure between the cylinder and the closed end of the tube. Atmospheric pressure would then push the piston back along the cylinder toward the closed end of the tube. In Papin's design, the tube played the part of the boiler, cylinder, and condenser. It was not a practical design and was only used for demonstration. As we will see, the evolution of steam engine design is primarily the separation of these three functions.

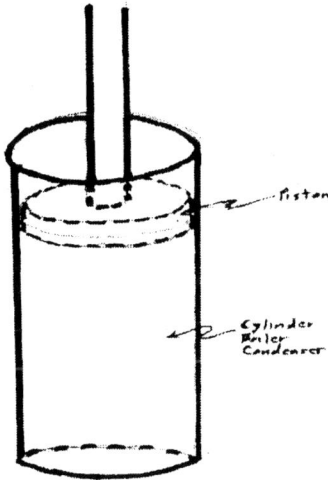

Fig. 2.3. Papin's first steam engine. The lower cylinder was heated, boiling the water in the lower part of the cylinder. It was then cooled by pouring water over the cylinder, condensing the water, and allowing air pressure from the top to force the cylinder downward.

The next step in the advancement of steam engine/pump design was taken by Thomas Savery in 1698. In Savery's model, there was a separate boiler for the water. Once the water had turned to steam a valve was opened and steam was introduced into a large container, driving out most of the air. The valve was then closed and the container was cooled, again using cold water, condensing the steam back to water and producing a low-pressure area inside the container. Another valve was opened that was connected to a pipe that led down to the water in the mine. Atmospheric pressure would force the water up the pipe into the low-pressure container much like it forces a fluid through a straw into a low-pressure area created in an individual's mouth. Here we see the separation of the boiler and condenser but no cylinder and piston. It was not really a steam engine because of the lack of the piston to do mechanical work. The design could only be used for pumping. Savery patented the design and tried to sell it as the "Miner's Friend." Because of its inefficiencies, due partially to the design and partially to the poor construction materials available at the time, the Savery pump did not sell well and was used very little. The patent, however, was valid for any machine that used fire to produce mechanical work.

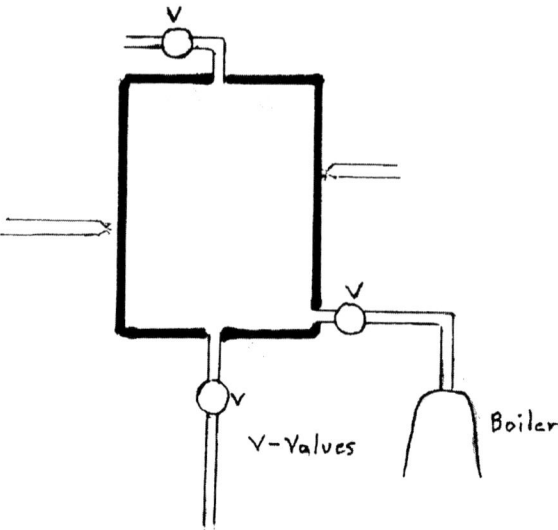

Fig. 2.4. Savory design. Steam was heated in the boiler to the right. The valves at the right and top were opened and steam entered the chamber, forcing any water out the top. The top and right valves were closed, and cold water was sprayed on the chamber through the tubes on right and left. Water condensed in the chamber, creating a low-pressure area. The bottom valve was opened, and water from the mine was forced into the chamber and the valve was closed. The process was repeated, forcing the water out the top and introducing new steam into the chamber.

Credit for developing the first practical steam engine usually goes to Thomas Newcomen. Although Newcomen appears to have developed the design of his steam engine independently, his machine was much like a working version of Papin's engine. His device was much larger than Papin's, had a separate boiler in which to convert the water to steam, and the piston was connected to a large wooden crossbar on a pivot. Once the steam was let into the bottom of the cylinder, the piston would move upward, moving one end of the crossbar up and the other end down much like the teeter-totter on children's playgrounds. The cylinder was then cooled, creating the partial vacuum and allowing atmospheric pressure to push the cylinder back down. This created an up-and-down motion on the crossbar, which was connected to a large piston pump that removed water from the mines.

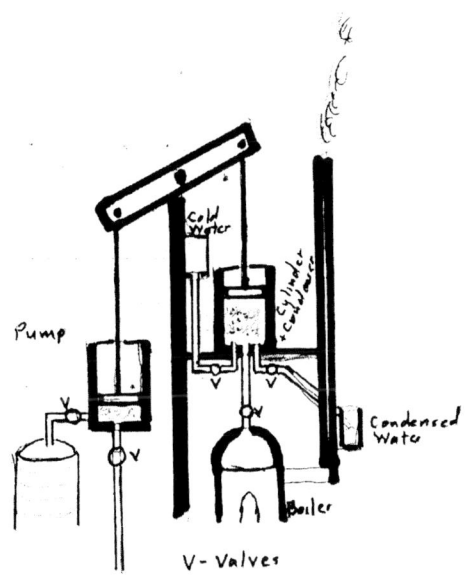

Fig 2.5. Newcomen's design. Water was heated in the boiler and a valve was opened allowing the steam to be introduced into a cylinder, which forced up a piston connected to a crossarm. The valve was closed and another opened, forcing cold water into the cylinder, condensing the water, and forming a low-pressure area. Outside pressure forced the piston down, moving the crossarm down. The process was repeated, giving the crossarm an up-and-down motion that was connected to a pump that removed water from the mine.

The first Newcomen engine was placed in a coalmine in 1712. That year could be considered the official beginning of the Industrial Revolution. The primary problem for Newcomen was that Savory held the patent on all engines that used fire to produce mechanical work. Newcomen was forced to go into business with Savery. Newcomen's "Fire Engine" was terribly inefficient, at about a 1 percent conversion rate of energy put into the machine verses work coming out. Because it was too big, unyielding, inefficient, and had a *huge* appetite for the consumption of coal, the Newcomen engine could not be moved or used any place but in the pumping of water at the mouth of coalmines. Even with those drawbacks the design ruled the pumping world for more than half a century. It could do the work of fifty horses, and there was no more reliable or efficient pump on the market.

James Watt, the son of a Scottish carpenter, was a sickly child who was not able to participate in normal childhood activities. He became adept at mathematics and making mathematical instruments. Around

1765 Watt became an instrument maker at Glasgow University and was put in charge of maintaining a small model of a Newcomen engine. Watt quickly realized the tremendous inefficiency of the Newcomen engine was due to the cylinder and the condenser being the same component of the machine. The cylinder was cooled when used as the condenser and would then need to be reheated when used as the cylinder. If he could separate these two functions, the efficiency of the engine could be vastly improved. For Watt seeing the problem was quite simple while designing a solution proved to be very difficult. After struggling with his "invention" for over a decade and working with several backers, Watt finally installed his first two engines in 1776, one to pump water from a coalmine and one to blow the bellows in an iron foundry. The machine that would power the Industrial Revolution had arrived. By separating the condenser from the cylinder, Watt was able to get four times as much power for the same amount of coal than the Newcomen design. Through other design changes, he was also able to give it a greater power to weight ratio and make it more reliable.

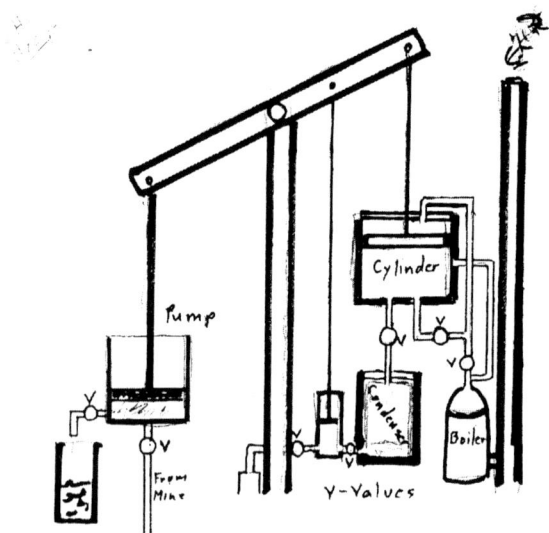

Fig. 2.6. The Watt design. Very similar to the Newcomen design, except there was a separate condenser that did not have to be heated and cooled on every cycle. A valve was opened after the piston was forced up and steam was forced into the condenser where it was condensed, and the piston was forced downward. Basic design would last for over a hundred years.

Although the machine to power the revolution had arrived, the problem of producing enough raw materials to construct the machines was still looking for a solution. The production of iron needed charcoal (pure carbon) to both supply the heat and drive off the oxides in the form of carbon dioxide from the iron ore. Since wood was in short supply in England, most of the iron was being imported from more wood-rich countries. The obvious solution was to use coal as a substitute for charcoal, but the impurities in the coal would be introduced into the iron, making for lousy iron. It would take more than a century of experimenting to learn how to bake the coal to drive off the impurities. Finally, in the early 1700s, cast iron was made using the baked coal (coke) and after another half-century of experiments, in the 1780s, wrought iron was first produced using coke.

Everything was in place for the Industrial Revolution: the engine that would power the revolution, the process for producing the material that would build those engines, and the fuel that would supply the energy needed. Between 1780 and 1830, the world changed. Mass production, mass transportation, and industrialization with the growth of the factory system all became staples of modern society. England would become the manufacturing center of the world and the dominant world power, both economically and militarily. Other countries would try to follow England's example but would lag behind by about fifty years. There were many problems associated with this new world, but they were largely ignored because of the promises of the time.

Manchester was the manufacturing capital of England and the world, with all of the associated promises and ills. The city boasted enough cotton mills to clothe a large part of the "civilized" world with cotton being fed to the mills from the slave plantations of the American South. A new type of civilization was being created, one that included an extremely wealthy industrial class, an expanded middle class to support the industrial giant, and an expanded labor force to serve that industrial machine. The industrial civilization would form its own values, belief system, and a greater separation between humans and the rest of nature. The perception became one of conquering nature instead of trying to work with nature. James Watt stated, "Nature can be conquered if we can but find her weak side." The perception of the children, men, and women of the industrial labor force by the middle and industrial classes became one of subhumans who could be crucified to the gods of industry and profit. In Manchester more than 57 percent of the children died before the age of five, rickets becoming endemic because of the loss of sunlight and vitamin D. The poor working class of Manchester had an average lifespan of only seventeen years, while the wealthy had a lifespan of thirty-eight years. Rural poor at that time had an average lifespan of thirty-eight, while the rural rich had a lifespan of fifty-two. It was all accepted in the name of industrial progress. Many, if not all, of these values and perceptions were carried along with the industrialization of the rest of Europe and the United States. Modern humans are still wrestling with the perceptions formed in nineteenth-century England.

Although Watt continued to make a number of improvements in his steam engine and patented his separate condenser design, he never continued improvement in his weight-to-power ratio because he failed to use high-pressure steam. Watt was aware of the advantages of incorporating high-pressure steam into his engine but felt it was too dangerous with the materials available at the time. Richard Trevithick in England and Oliver Evens in the United States were also aware that the use of high-pressure steam would lead to higher speeds, better power-to-weight ratio, and a wider expansion stroke. Watt's patents had expired by the early 1800s, so Trevithick built the first high-pressure steam engine using a design similar to Watt's, which he called the "Cornish Engine." It became a great success in pumping water from mines and for other industrial needs in England after 1800, but perhaps its most innovative use was Trevithick designing the first steam-powered locomotive in 1804.

By the early 1800s, the horse-powered railroads had become popular across England because of the tremendous mud problem throughout the British countryside. Wooden tracks had been used since the Middle Ages, and metal tracks were first put to use in transporting coal from the mines to the cities after coal became a primary heat source. In 1804 Richard Trevithick designed a steam locomotive that used his Cornish engine as a power source. The idea worked fairly well, and Trevithick was able to transport heavy loads at a then reasonable speed, twenty-five tons at slightly less than five miles per hour. In his design, Trevithick inserted a vertical chimney that drew the exhaust steam past the boiler creating, much greater draw through the boiler. It was not a commercial success because of high failure rates, inability to maintain the high steam pressures needed, and being so heavy as to break the rails. Trevithick also designed steamboats, steam carriages, and steam shovels, none of which became a commercial success.

Although Trevithick was never able to successfully implement any of his designs on a commercial basis, the fundamental ideas regarding steam transportation were ready for others to run with. In 1825 George Stevenson designed and built the first practical railroad to run twenty-six miles between the coal-mining town of Darlington and the river town of Stockton. In 1830 the first public railway was opened between Liverpool and Manchester. It was an instant success, and soon railroads were popping up all over the world. In the United States, in 1807, Robert Fulton and his partner, Robert Livingston, built and successfully demonstrated the first steamboat with a trip up the Hudson from New York to Albany. Soon steamboats were on virtually every river in the United States and Europe. The era of steam transportation had arrived and would transform the world.

During this time of tremendous invention and technical progress, science in the study of energy and power was also progressing. In 1824 a young French genius, at the age of twenty-eight, published theoretical research that would revolutionize the study of thermodynamics. The young man was Sadi Carnot, and the treatise was entitled the "Reflections on the Motive Power of Heat and Machines Fitted to Develop This Power." In his treatise, Carnot addressed two questions: First, is there an upper limit to the amount of work that can be developed with a heat engine? And second, are there methods other than the traditional steam engine to achieve this work? He introduced the concept of the Carnot engine, which is an ideal heat engine, and showed that the maximum efficiency an engine can achieve is dependent only on the temperature difference between the gaseous fuel entering the engine and the burnt gases leaving the engine. In the paper, he also presupposed the idea of the gas turbine that would be the next major step in steam-power design. Carnot died at the early age of thirty-six. His unfinished papers show he was decades ahead of others in understanding the thermodynamics of heat engines.

Carnot's work showed the existing piston-driven steam engines with external boiler were, by design, a quite inefficient power source, anywhere from around 5 percent to 10 percent. Attempts had been made to increase that efficiency, but Carnot's work had shown that with the existing materials and designs, that range was the maximum attainable. Hero's Alexandrian steam engine had demonstrated that expanding steam coming out of an orifice (jet) could do work. In 1837 an American, Avery, in New York and an Englishman, Greenock, had used the same concept for driving circular saws. Avery built a five-foot-diameter tube with jets at each end connected to a rotating saw. Steam was introduced into the tube, exiting through the jets, rotating the saw at speeds approaching nine-hundred feet per second. The wheels were quite inefficient and never a commercial success; however, by the late 1800s, the science of thermodynamics was becoming a powerful tool, and mathematics was showing the energy in expanding gases could be used quite efficiently. Nearly 80 percent of the energy in the expanding gases was manifested in the form of molecular velocity.

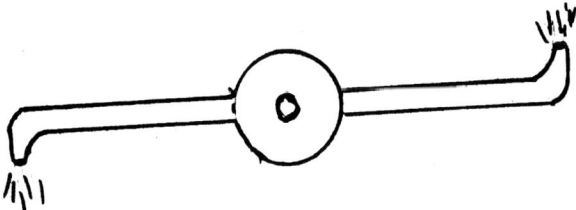

Fig. 2.7. Avery's first steam turbine. Steam was forced from the center out through the blades, where the escaping steam produced a rotational motion in the blades.

Using the concept of extracting energy directly from the moving molecules, a Swedish scientist, Dr. de Laval, made the first steam-powered turbine by forcing expanding steam to pass over a metal paddle wheel. Again, the idea was innovative, but the rotational speed of the paddle wheel would have had to exceed the structural limitations of the material. The centrifugal forces involved at the rapid rotation required were too great for any practical commercial design.

In 1884 Sir Charles Parsons of England was able to split the amount of energy (and therefore velocity) imparted to each wheel by designing a turbine that had several wheels in series. Each wheel would receive less energy and would then rotate at slower speeds, allowing for lower rotational forces. The British marine industry and Navy saw the promise in the Parson's turbine design and invested large sums of money to develop a practical machine. By the turn of the century, virtually every ship in the British Navy was powered by a steam turbine and in 1900, it was first used for powering electrical generators in the United States. Parson's design, with improvements in rotor technology and structural material, is the same basic design in use today. It is very simply the most efficient and practical heat engine ever designed.

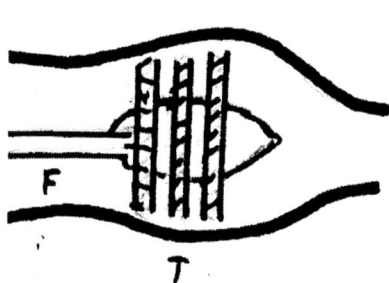

Fig. 2.8. Parson's first practical steam turbine. In this design, water is heated and the expanding steam is passed into a high-pressure area (F), then over a series of rotor blades (T), into a low-pressure area behind the blades, forcing them to rotate. Steam exits the back.

Over the next twenty to thirty years, coal continued to be the fuel of choice to boil water to generate steam to power the turbine engine. Ultimately in both England and the United States, the pollution effects of coal became too great a factor to ignore, and by the 1960s, most electrical generation had been converted to natural gas or petroleum. As natural gas and petroleum become more expensive with diminished reserves, new combustion technologies are being investigated in an attempt to clean up the pollution effects from coal. One of these technologies is termed "staged combustion," which combines more oxygen during combustion, thereby reducing the nitrous oxides produced by 90 percent. Another new technology is fluidized bed combustion, where the coal is pulverized into a powder and burned in suspension. In this technique, lime can be added, which acts like a sulfur sponge to absorb the sulfur pollutants. To eliminate other pollutants, primarily the heavy-metal poisons, some are promoting pre-baking the coal to drive off those pollutants much like coke is produced by heating to drive off the impurities. All of these pollutants and combustion technologies will be examined in chapter six.

Whether coal plays a large part in our energy future is still being debated. As we will see in chapter six, even with the modern technologies, the combustion of coal will always release large amounts of carbon dioxide into the atmosphere, contributing to global climate change. In fact, coal produces more carbon per energy unit generated than any other fossil fuel. Most environmentalists feel that with the overwhelming evidence of global climate change occurring that no new power plants should be installed without the technical processes available to sequester the carbon released in the generating of electricity.

The political and economic forces behind the promotion of a coal economy are very powerful, however, and seem to be more interested in maintaining that power than in promoting a healthful environment for our planet. With the rush to put more coal-fired plants on line, the new technologies are being labeled as too expensive and not being incorporated in the new designs. Virtually all coal-fired electric generating plants are still using outdated technology and strongly resisting the upgrading to more modern and cleaner combustion methods. At this time, however, because of the political and economic power of the coal enthusiasts, it looks like coal will play a significant part in our immediate energy needs.

Chapter Three

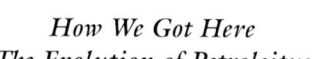

How We Got Here
The Evolution of Petroleitus

Petroleitus: The end product of the evolution of the human species into a branch that is completely dependent upon petroleum.

Author's Note: This chapter briefly discusses the growth of the world petroleum industry. For a more detailed history, there are several texts that offer a serious discussion of that growth. Perhaps the book that most vividly and thoroughly chronicles the development of the world petroleum industry is *The Prize*, written by Daniel Yergin.

Petroleum: In the Beginning
One would like to think there is a rational and evolutionary growth to science and technology. Not so. It appears that many of the developments in these two areas have similar motivations, brilliance, curiosity, search for economic security, greed, search for esteem and, quite often, just blind luck. The development of the petroleum industry has had its share of all of these.

Petroleum is certainly not a modern discovery. Early humans had seen this black goo on the surface in several areas of the planet, and over the centuries, found various applications for that goo. It was primarily used as a medical balm, but they also found uses as a lubricant, light source, building mortar, caulking, and an ancient weapon (Greek fire). Since antiquity the removal process was either soaking the oil in a rag and squeezing it out or digging a pit and, with a bucket, removing the oil from the pit. During the nineteenth century, Europeans learned to refine the oil into various products by simply heating it. The products, including naphtha, gasoline, kerosene, petroleum jelly, fuel oil, and lubricants, will evaporate quite easily.

The products will condense at different temperatures and are separated (distilled) during condensation. They also found that kerosene (initially made from coal oil) could be used as a source of light if burned in a proper lamp. Because of the recovery process, however, it was not readily available and was expensive. To make the kerosene, distilled from petroleum, more available to the general public would require an efficient removal process.

The oil and natural gas industries got their start as an outgrowth of the salt industry. Fifteen hundred years ago, the Chinese developed a method for retrieving salt from below the earth's surface using a drilling derrick. The technique spread to Europe and later the United States. Quite often salt producers would find oil and natural gas while they were drilling for salt. The byproducts were considered nuisances and would be disposed of in nearby rivers or simply dumped. In the early 1800s, in the eastern United States, salt producers found that petroleum and natural gas had commercial value and were sold locally.

The individual who would initially use a drilling derrick in the United States for the sole purpose of retrieving petroleum was George Bissel. He would later be called the father of the world petroleum industry. (In fact, the Europeans had been using this technique for a decade for retrieving petroleum from near the Caspian Sea.) Bissell was an extremely intelligent individual who could never seem to find his calling. He graduated from Dartmouth College and went on to become a writer, high school principle, college professor, and journalist. He seemed interested in everything and could speak several languages. His journey in becoming the "father" of oil production began in 1853 when, as an unemployed journalist, he was passing through western Pennsylvania to visit a professor friend at Dartmouth. He noticed workers in a field skimming oil with rags and found they were extracting petroleum. When arriving at the friend's office, he noticed a jar filled with the same rock oil (the name given to petroleum in that area) and wondered if it could be used as an illuminant. Bissel formed a group of investors who hired a chemist to analyze the oil. The chemist, Benjamin Silliman Jr., reported in 1855 that the rock oil could be brought to various temperatures and distilled into several products, one of which was a high-quality illuminant for which the investors hoped.

The problem, as always, was how to get the oil from the ground. It was already known that digging and skimming would not supply enough to be profitable. Bissel was in New York trying to figure a better way to remove petroleum from the ground when he passed a window with an advertisement for rock oil medicine. The advertisement showed drilling derricks in the background and said the rock oil was a byproduct of the salt-boring process. Bissell's epiphany was to adopt the drilling process directly to the recovery of petroleum.

He convinced the investment group of his plan, and they determined to go ahead with the drilling process on a farm near what is now called Oil

Creek, outside of Titusville, Pennsylvania. One of the members of the investment group, James Townsend, a New Haven banker, roomed in a hotel where a jack-of-all-trades and out-of-work conductor, Edwin Drake, also roomed. They became friends, and Townsend decided Drake would be the person to manage the drilling operation. Before sending Drake to Titusville, Townsend felt he should have a title to impress the locals. So was born "Colonel" Edwin L. Drake.

It appears Townsend and the investment group had the right person for the right job. In spite of overwhelming difficulties and just as the investment group ran out of financing, on August 27, 1859, at the depth of sixty-nine feet, Drake struck oil. It was not a gusher as the movies have made it, but Drake was able to attach a hand pump and draw up the black liquid. A mad rush to acquire land followed. The population of Titusville escalated overnight, and land prices jumped. The petroleum industry had arrived.

Drake's discovery led to the first oil boom with the accompanying escalation in land prices. Within fifteen months of Drake" discovery, there were seventy-five producing wells. Cheap refineries, to turn the oil into kerosene, were set up in the oil regions and in Pittsburgh. Originally the oil was transferred from the well to the refineries by horse-drawn wagons operated by Teamsters. The oil was stored and shipped in barrels. The Teamsters charged exorbitant prices, and it cost more to ship the barrels a few miles to the railroad than the shipping cost from Pennsylvania to New York by rail. The first pipelines (originally made from wood) were developed to cut these shipping costs. Later tank cars were developed for the railroad, and barrels were no longer used. We still, however, use the barrel as the standard unit of volume when describing an amount of oil.

Fig 3.1. Edwin Drake (right) in front of his oil derrick near Titusville, Pennsylvania.

Over the next fifteen years, the petroleum industry grew rapidly. Fortunes were made and lost overnight, and there were booms and busts, under-production and over-production. Land going for a million dollars an acre during a boom would sell for fewer than five dollars an acre when the oil ran out. There was no standardization on the refining process, and thousands of people died in fires because of too much naphtha or gasoline in the kerosene. It was an extremely chaotic time in the petroleum industry.

Out of the Chaos Came Standard

Perhaps no name or company is more synonymous with the growth of the oil industry than John D. Rockefeller and the company he created: Standard Oil. The modern oil industry really began in 1865, when Rockefeller was twenty-six. He bought out his partner in a refinery in Cleveland and started his own company. He was, arguably, the greatest business genius ever produced by the United States. Rockefeller seemed to understand what the nation wanted and needed. He could foresee the long-term direction of the company and still keep a firm hand on the day-today operations.

John D. was a very orderly person and didn't like the frenzied, chaotic activities surrounding the infant oil industry. To eliminate the chaos and impose his brand of order, Rockefeller and a group of investors started developing or buying refineries and pipelines. He wanted to standardize the industry and therefore chose the name Standard. Over the next fifteen years, with the use of threats, pricing wars, open deals, and shady deals, John D. and his partners gained ownership of nearly all of the pipelines and refineries in the U.S. By the end of the 1870s, through their refineries and pipelines, Standard Oil had a virtual lock on oil production in the United States. In the process, they made enemies and destroyed careers. Later this would come back to haunt them.

Standard Oil virtually designed the modern oil industry. They were the first to introduce scientific and technological research into oil production and refining. With that research came an extremely reliable product. They simplified oil transportation with pipelines and rail tankers and developed an extensive marketing system. Rockefeller developed a "trust"—a conglomeration of refineries and pipelines that appeared to be independent but were actually controlled by Standard. Through their control of refineries and pipelines, they could control the price paid for the oil at the wells. Standard Oil became the most advanced business monopoly of their time in the United States, and they began spreading the operation overseas.

Until 1885 Standard Oil stayed out of the production area and concentrated on the peripherals: storage, pipelines, and refining. Rockefeller, in his quest for order, felt exploration and production was too risky and that others should take that risk. In 1885 the state geologist in Pennsylvania warned that the oil regions of Pennsylvania would soon be depleted. This caused concern for Rockefeller. With no oil, storage, pipelines, and refineries would have no purpose. The same time the oil fields were becoming depleted in Pennsylvania, a major discovery of oil was made in Ohio, then in Indiana. Rockefeller was determined to buy all of the leases available to guarantee a supply of petroleum. He would impose his brand of order on production. By 1891 Standard was responsible for a quarter of all oil production in the U.S.

Standard Oil began marketing its products in Europe and Asia and looking for sources of oil in other places in the world. While the infant oil industry was getting its start in the U.S., the Europeans could also see the need for kerosene and the possibility of massive profits. The first European source for oil was found in Baku, on the Caspian Sea in Russia (what is now Azerbaijan). It had been know as a source of oil for several hundred years, but the method of production was digging pits and skimming. Although the world's first oil-drilling rig was placed there around 1849, the Nobel brothers did not commercially developed the area for another twenty-five years. (Alfred Nobel of the "Nobel Prize" fame was one of the three.) They began producing oil in 1874 and, within ten years, produced almost eleven million barrels, nearly a third of American production. Initially transportation was such a problem that they could only sell the oil locally. In most of Russia, it was cheaper to import Standard Oil products from America, eight thousand miles away, than from Baku, less than four hundred miles away. One of the Nobel Brothers, Ludwig, designed the first oil tanker to help eliminate this problem. The brothers were then able to ship their products over most of Russia.

Fig. 3.2. The Baku area of central Asia. Baku is on the west bank of the Caspian Sea.

Because of the limited market for the oil products in Russia and the very large supply of oil in Baku, investors tried to build a railroad from Baku to the Black Sea. The railroad would open up the Baku supply to the rest of Europe. Finally, with the backing of the French Rothschilds, the railroad was completed and Baku oil was available for the European market. The Rothschilds formed the Russian and Black Sea Petroleum Company, now know by its Russian Initials, "Bnito." After the Nobel Company, it became the second largest producer of Russian Oil.

Ultimately the Rothschilds needed transportation for Bnito products and would form a business partnership with Samuel Marcus, a London shipping broker. Marcus commissioned the building of newer, larger, and technologically advanced oil tankers and would go on to develop the largest fleet of oil tankers in the world. Marcus called his company Shell Oil, after his father, who was a seashell merchant.

The second largest European supplier of oil, and competitor for Standard, was a Dutch group that found and developed oil in the Dutch East Indies. The area is now known as the island of Sumatra, in Indonesia. The company, known as Royal Dutch, was founded in 1890 and was initially under-financed. With new financing, supplied primarily by the British, the company ultimately became a thriving, primarily British, entity known sometimes as "British Dutch." By the turn of the century, Shell and Royal Dutch controlled over half of the oil exported from Russia and the East Indies and, along with Standard, controlled 80 percent of world petroleum production. The competition among the three was extremely tight, with Standard constantly trying to ruin or buy out the other two. Because Shell was taking the worst beating, Marcus (with his dislike for Standard) decided to cast the fortunes of Shell with Royal Dutch. The new company was known as the Royal Dutch/Shell group. Here at last was a major worldwide competitor for Standard Oil. Royal Dutch/Shell had all of the financial, technical, petroleum, storage, transportation, and human resources that Standard could muster.

Standard Oil watched this expansion of Russian Oil into the European market with alarm. They saw competitors were cutting into their worldwide market. Standard initially tried price cutting and when that didn't work, they went to their bag of dirty tricks. They started rumors about accidents caused by Russian Oil and tried sabotage and bribery—all tried-and-tested techniques that worked in the U.S. The Nobels and Rothschilds would not quit and fought back successfully. Finally Standard formed a European company, Anglo-American Oil Company, which joined in ventures conducted by local producers. The American share of the world oil market would fall from 78 percent to 50 percent over the next few decades.

While the world was waking up to the oil age, the boom was proceeding in the U.S. After Indiana, oil was discovered in California. The initial supply was not large and with the distance involved, California oil was not initially considered a valid source for the populated Eastern states. On January 10, 1901, what was thought to be the motherload of all oil sources was discovered on a hill known as Spindletop outside of Beaumont, in southeast Texas. The developers were originally hoping for fifty barrels a day. They got a gusher that supplied an amazing seventy-five thousand barrels a day. Although Spindletop, as a source of oil, lasted only three years, it was such a prolific source that it changed perceptions as to the amount of oil that was thought to be available as a fuel that would power the new technologies and our modern civilization.

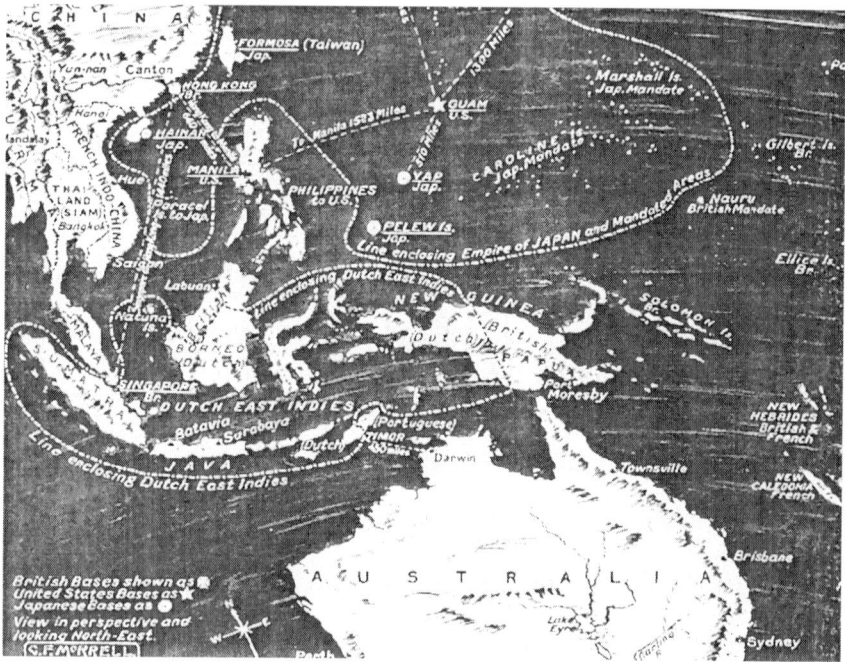

Fig. 3.3. Oil was discovered in the Dutch East Indies (now Sumatra and part of Indonesia).

Standard felt that Texans had a bad attitude toward John D. Rockefeller and were quite uncivilized. As a result, they did not originally take advantage of the Texas and Oklahoma oil strikes. Gulf Oil was formed by the backers of the drilling at Spindletop and later built a pipeline across Oklahoma and Texas. Also, Sun Oil and Texaco were formed as a result of the Texas and Oklahoma oil strikes. Standard's share of American oil production was falling, from a high of 90 percent in the late 1800s to about 60 percent in 1911.

Standard was also facing problems on the political and legal front. Over the past quarter-century, Standard had risen to the top through price cutting, starting rumors, devious deals, buying political favor, and an assortment of other questionable practices. They were very secretive about their operations and were sometimes blamed for events with which they had nothing to do. Also, the turn of the century brought a very liberal, progressive view to American society and politics. The reform movement included unions, improved working conditions, and the examining of business practices. The force driving much of the new attitude was the press. In this atmosphere, Standard and John D. Rockefeller would soon come under scrutiny.

The *McClure Magazine* hired one of the first female journalists, Ida Minerva Tarbell, to write an investigative report on Standard Oil and their business practices. Ms. Tarbell was a thorough investigator and top journalist, and her father's career had been destroyed by Standard. Her series began appearing in *McClure's* in November 1901. It was a bombshell. For twenty-four months, the articles very thoroughly described all of Standard's practices in gaining their monopoly. In 1904 the articles were published as a book and could possibly be called the most influential business/political writing of the twentieth century. Tarbell felt there should be some counterforce to curtail the power of growing monopolies. Teddy Roosevelt, who took office in 1901 after the assassination of William McKinley, felt the government should be that force.

Roosevelt's administration started an investigation into Standard Oil and, in 1906, brought suit against Standard Oil under the Sherman Antitrust Act. In 1909 the federal court found in favor of the government and ordered the breakup of Standard. The company was divided into Standard of New Jersey (Exxon), Standard Oil of New York (Mobil), Standard Oil of California (Chevron), Standard Oil of Ohio (Sohio and later the American arm of British Petroleum), Standard Oil of Indiana (Amoco), Conoco, and Arco. All but a few of American oil companies had their start as Standard.

The dissection of Standard turned out to be a very beneficial surgery for the owners, for the company, and for the world. The owners became even wealthier with their interest in all of the companies formed. The individual companies were allowed to move ahead on their own initiative, and all became successful. The world gained because of the increased amount of research by each company developing new and better products. The research would be needed in a changing world where old markets were disappearing and new ones were appearing.

The Light and the Engine

Until the late nineteenth century, the world petroleum industry was based primarily on kerosene. In the 1870s, within a three-year period, two developments would take place that would radically change the basis for the industry. The first occurred in 1876, when a German engineer, Nikolaus

Otto, developed the first prototype internal combustion engine. The engine was based on a theoretical design proposed by Frenchman Alphonse Beaude Rochas in 1862. The manufacturing of the engine in the United States began in 1878. It was soon found that gasoline, a byproduct of the manufacture of kerosene, was the ideal fuel for the internal combustion engine. Gasoline had formerly been sold at any price the producers could get and was quite often disposed of illegally by being dumped into waterways. Also, in the late 1800s, engineers found that fuel oil could be used more efficiently than coal in powering the large steam engines and steam turbines that were propelling the ships of that era. In several of the navies of that time, the smaller ships were being converted to fuel oil.

The second development occurred in 1879, when Thomas A. Edison developed the first incandescent lightbulb. He built a demonstration project in New York City and flipped the switch from his banker's office in 1882. So began the United States Electrical Energy Grid. By 1902 18 million lightbulbs were in use. Electric lighting was safer, brighter, more convenient, and cheaper than the kerosene lantern. It soon brightened most of the larger cities in the Western Hemisphere. The growth, application, and extent of electrical energy use will be discussed in the next chapter.

From the beginning of its development, there was very little doubt that in time, the electric light would completely replace the kerosene lantern. It was not so obvious the horseless carriage would replace the horse. In the late 1880s and early 1890s, several inventive individuals began installing the internal combustion engines into a buggy, and the automobile was born. The first auto race was held in France with the impressive speed of fifteen miles per hour being attained. The next year, the first U.S. auto race was held and was much slower and much more boring. It was here that the immortal phrase "Get a horse" was coined.

Besides the internal combustion engine, the horse was being challenged by the electric motor and steam engine. Over the next few decades, however, the muffler and more efficient engines were developed, and with the ability to easily carry large amounts of gasoline fuel, the internal combustion engine won out. It is easy to wonder what would be the state of the world transportation system if the internal combustion engine had not so overwhelmed the electric car. We would probably have much more efficient electric storage methods that would give much greater range and power to the electric vehicles. The world is just now beginning to develop these larger, more efficient storage methods. These will be discussed in chapter four.

About twenty-five years after the development of the first internal combustion engine, Rudolph Diesel developed the engine that bears his name. He originally tried to develop an engine that could burn coal dust but ended up with the heavier, more efficient liquid fuel engine. The new engine burned fuel oil and would be used to power trucks, trains, and ships. It was extremely reliable and could also run on other types of fuel, including vegetable oil.

The benefits of a world running on oil soon became obvious, first to engineers and later to politicians and the public. To make the change from horsepower to petroleum power, however, would require confidence in a continuous, long-term supply of oil (gasoline). The impressive strike at Spindletop convinced a large part of the public that the supply was available. Another hindrance was that the refining process of petroleum yielded only about 15 percent to 18 percent gasoline. With the increase in research after the breakup of Standard Oil, Standard of Indiana developed the cracking process that could yield up to 45 percent gasoline. Everything was in place for the transfer to a worldwide petroleum economy. Twentieth-century humans would evolve to become "Petroleitus."

The Fruits of War?

For better or worse, nothing changes perceptions and modes of operations quite as quickly as war. So was the case with the acceptance, and ultimately dependence on petroleum. Previous to 1914, all battleships used coal as a power source. All of the world nations used horses to move men and artillery. By the end of the First World War, all the world navies were either powered by or being converted to fuel oil. Tanks, trucks, and automobiles were being used to move men, artillery, and now fuel. Airplanes powered by the internal combustion engine were used, first for observing and later as weapons of war. Future battles and wars would be fought for the protection of oil and rights to oil fields. Winston Churchill, when appointed to First Lord of the Admiralty in 1911, became convinced of the benefits of petroleum and committed the British navy to fuel oil. He ordered the first steam turbine, fuel oil-powered battleship. He also saw the problems associated with that dependence, in the amount of men and blood that would be needed to protect the world's sources of fuel oil. For all practical purposes, following World War I, the world was dependent on petroleum.

The Petroleum World

With cars, trucks, planes, and ships all running on gasoline, there was a tremendous increase in the use of petroleum from 1917 to 1920—so much so that several U.S. geologists proclaimed that with the increased demand, there would soon be an end to U.S. oil production. Technology, however, would come to the rescue.

In the early search for oil, searchers were limited to the surface of the planet and would look for various geological structures that appeared to indicate underground oil reservoirs. Technology developed during the war would lead to the ability to get a peek underground. The first of the new techniques was the torsion balance. It was an extremely sensitive balance that could determine changes in density near the Earth's crust. The changes in density could indicate the presence of an oil field. The second development was an extremely sensitive magnometer which could indicate changes in the Earth's magnetic field that may also indicate oil. The most powerful tool was the development of the seismograph. Dynamite was set off, and the shock wave would travel through the earth. The wave would be either refracted (bent) or reflected by underground structures that could contain oil. With the airplane and aerial photography, a much larger area could be surveyed and became an important tool of the oil hunters.

With the new tools, oil was discovered in several places around the globe. In the U.S., major fields were opened around Los Angeles, in Oklahoma, and Texas. Mexico and Venezuela, in Latin America, had major discoveries as well. When one of the major discoveries occurred, the same chaos would surround the opening of new wells that occurred in Rockefeller's time. Everybody and their distant relatives would try to put up a derrick. In California the derricks were placed so close together that their legs would overlap. The discoveries and following chaos led to two events during the 1920s.

First, there was a tremendous glut of oil and the price fell drastically. Second, several individuals were starting to see the overall harm this chaotic method of drilling was bringing to the industry and supply of oil in the U.S. These individuals were convinced that the large number of wells being drilled into a reservoir would quickly lower the pressure in that reservoir. As the oil reservoir lost pressure, the oil would not be forced up the pipe and the well would appear dry. This was exactly what was happening. The theory, however, was not accepted by most oil people, and it was virtually impossible to get the oil industry to cooperate and develop the fields in a more organized manner. In 1924 President Coolidge formed the Federal Oil Conservation Board to investigate the problems in the oil industry. Investigate was all the board did, and the decade of the 1920s, as in many other areas of American society, brought overproduction and waste to the oil industry.

Also during the twenties, the large oil companies formed the American Petroleum Institute to promote their political agenda and company propaganda. Their political agenda was simply no government interference and no tariff on Mexican and Venezuelan oil, most of which they owned. Small oil producers wanted a tariff on foreign oil and formed the Independent Petroleum Association to promote their political agenda.

In the mid-twenties, the first gas stations opened that were owned by one of the large oil companies: Phillips Petroleum. Soon all of the major petroleum companies followed suit. Now the large companies would control all operations in the oil industry, from production, refining, storage, pipelines, shipping, and retailing.

On October 3, 1930, the largest oil strike ever located in the United States was found in East Texas. By June 30, 1931, one thousand wells had been dug and East Texas was producing five hundred thousand barrels of oil per day. The bottom fell out of the market. There was complete panic from the large producers on down. Even though the production cost for oil was eighty cents a barrel, it was being sold off for as low as six to eight cents a barrel. Complete anarchy ruled the oil fields. Every producer tried to undercut other producers. There was no cooperation. Agreements were made and broken the next day. Pipelines were blown up. By 1933 even the most hardened oil anarchist could see that survival of the oil industry would require government intervention. The intervention came in the form of Franklin Roosevelt's "New Deal."

The result of the New Deal legislation was a Producer, Federal, and State government partnership. Prices were raised by controlling production and eliminating "bootleg" oil sales. Heavy import tariffs on foreign oil were imposed. The Interstate Oil Compact was formed, which helped states exchange information, standardize legislation, implement rationing, and encourage conservation. Prices increased to between a $1.00 a barrel and $1.18 a barrel. The oil industry had been saved. Later OPEC would form under similar guidelines to the ones established at this time.

Even with the world glut in oil, the thirties were spent developing new oil fields in different parts of the world. In spite of the world's leading geologists stating the Middle East would be "oil dry," most of the wells were developed in that area. Saudi Arabia, Kuwait, and Iraq were opened up to oil exploration and formed agreements with the various western oil companies. As war approached, the different governments tried to form agreements with the oil producing countries that would allow them access to the oil and deny access to their perceived enemies. War, however, would delay the development of Mid-East oil.

World War II

It could be argued that World War II was primarily fought over oil resources. There were certainly many other reasons for the start of the war. Even so, most of the battles were fought as a result of plans to gain access to the world petroleum reserves.

In Japan the military was united in developing Asia into a "spirit of coprosperity and coexistence based upon the Imperial way." Japan wanted control of Asia and control of Asia's resources. To accomplish this control, Japan, being poor in natural resources, needed a guaranteed supply of petroleum, primarily petroleum in the Dutch East Indies. To gain access to the

Dutch East Indies oil fields would require control of Thailand, Malaysia, and Singapore. The Japanese planned for the conquest of these countries but were concerned about the United States Pacific Fleet attacking their flank once the invasion of Southeast Asia began. In order to eliminate this concern, they felt they must cripple the U.S. fleet in Pear Harbor.

Fig. 3.4. Map of Asia. Shows distance Japanese tankers had to travel for East Indies' (Sumatra) petroleum.

Pre-war Japan received 60 percent of their oil from the United States and most of the remaining 40 percent from the Dutch East Indies. They had very limited oil reserves in the Japanese Islands and even tried to install a coal gasification industry. The plan was given up, as it was too time-consuming and expensive. During the thirties and early forties, Japan became more intent on their conquest of Asia, and the relationship they held with Europe and the United States began to deteriorate. Even so, the U.S. continued to supply Japan with oil until less than a year before the attack on Pearl Harbor.

The relationship between Japan and the United States had deteriorated to such a degree by late 1941 as to be nonexistent. It seems as though virtually all military and government leaders in the U.S. expected an attack from Japan by late November or early December. The U.S. had even broken the Japanese code and intercepted communications showing an increase in activity but could not interpret specific intent. The military command in Hawaii felt an attack on Pearl Harbor was impossible, and Japan could never attack the fleet while anchored there. The results are well documented in texts and film.

Although the extent of the destruction to the U.S. Pacific Fleet was more than Japan had ever hoped for, the attack was not as successful as it could have been for three reasons. First, the aircraft carriers were not in the harbor because of maneuvers and escaped without a scratch. The war in the Pacific would be become known as the "Carrier War." All of the major navel battles were won because of damage inflicted by navel aircraft based on those aircraft carriers that escaped damage at Pearl Harbor. Also, without those same aircraft flying cover, no invasion of the Pacific Islands would have been possible. The primary reason Japan lost the war in the Pacific was because of the aircraft carrier.

Second, the U.S. still had their submarine fleet. The submarines would prove decisive in denying the Japanese oil that had to be shipped from the East Indies. In their rapid conquest of Southeast Asia, the Japanese quickly overran the Dutch and British oil fields in Sumatra. The fields were destroyed when the British and Dutch left, but the Japanese quickly had them producing again. The problem was the distance and route the cargo ships had to pass through to get the oil back to Japan. The route made the ships particularly vulnerable to American submarines, which the Japanese did not foresee. In a herculean effort, after suffering through two years of incompetently designed torpedoes, 55 percent of Japan's total merchant shipping was destroyed by the U.S. submariners. Another 40 percent was destroyed by surface ships and aircraft. By late 1944, Japan had virtually no oil to power their navy and air force.

The third reason the attack on Pearl Harbor was not a definitive success was the attacking planes failed to destroy the naval oil reserves. The Navy had four and one-half million barrels of oil stored on Oahu. The Japanese failed to strike any of it. This was the same oil that would supply the aircraft carriers and submarines that inflicted so much damage on Japan. The Americans would have been helpless without the oil. One can only assume the Japanese navel planners did not yet realize how important the oil reserves were to a modern navy. They failed to send their planes back to destroy those reserves. In his memoirs, Admiral Halsey stated that if the Japanese had destroyed the oil supplies, the war in the Pacific would have been extended for at least two more years.

Germany also lacked a supply of petroleum within its own boundaries. To become self-sufficient in liquid fuels, they developed processes for making synthetic fuels out of coal. The first, the Bergius Process, sometime called hydrogenation, combined a large amount of hydrogen with coal under high temperature and pressure in the presence of a catalyst to produce a high-grade fuel. In another process, the Fischer-Tropsch process, coal molecules are broken down under steam in the presence of hydrogen and carbon monoxide. The products react together to form synthetic fuel. The Germans formed a synthetic fuel industry using the Bergius Process, which by 1940 supplied 45 percent of German oil and 95 percent of their aviation fuel.

The German company responsible for the research into synthetic fuel formed a partnership with Standard Oil of New Jersey. With the glut of oil on the American market, Standard found it was not economical to produce synthetic fuels. They did, however, find the same process could be applied to petroleum to produce a larger percentage of gasoline from a barrel of oil.

At the beginning of World War II, the German military had overwhelming and unexpected initial success with their blitzkrieg (lightning war) and quickly overran Western Europe. The only thing stopping Hitler's armies of complete domination of Europe was the British in the west and Russia in the east. Russia and Germany had formed a non-aggression pact, which Hitler had no intention of honoring. He hated Bolshevism and felt it was his life's duty to eliminate it. Even more importantly, he felt the German war machine and industry needed greater access to oil resources, which were located in the Soviet Union, and defeating the Soviets would be the solution to his problems. On June 21, 1941, Operation Barbarossa began. Of the many mistakes made by Hitler during the war years, the attack on Russia was certainly the greatest.

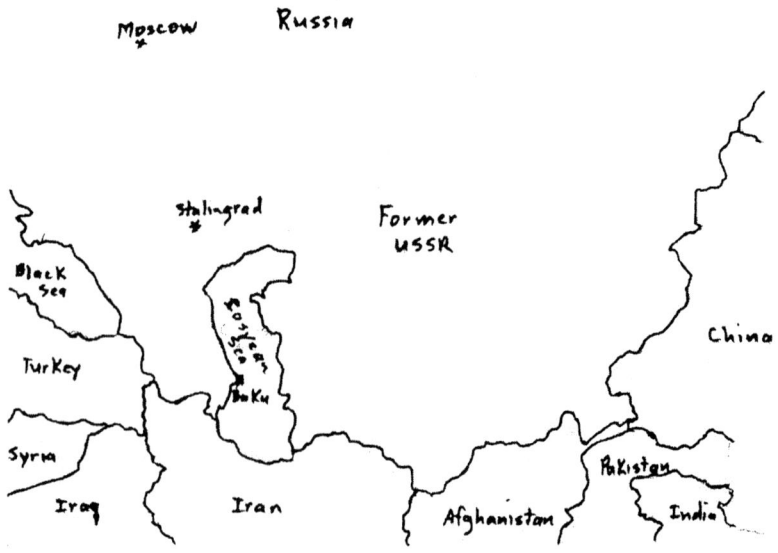

Fig 3.5. Germans attack Russia to gain access to the Baku oil fields. They were defeated at Stalingrad in one of the bloodiest battle of World War II.

The attack began well for the Germans—in fact, probably too well. They soon outran their supply lines and had to halt. The Russians were able to regroup and stop the attack. The Germans never did overrun the oil fields in the Caucasus, primarily Baku, and soon ran out of oil. The following battles resulted in an almost unimaginable loss of life to both sides, especially the Russians. The battles did, however, signal the beginning of the end for the German war machine.

In North Africa, the same scenario was happening to Rommel's Africa Korps. In early 1942, he began his attack on the British, and again with unexpected overwhelming initial success, Rommel outran his supply line. He was forced to halt at a small town in Northern Egypt called El Alamein. Both Rommel and Hitler felt it was just a matter of time before Egypt fell and the Afrika Korps would continue through the Middle East, around the eastern side of the Mediterranean Sea, and join up with the German army attacking Baku. They were to be proven wrong, again because of lack of oil supplies. The British still had naval control of the Mediterranean, and very few supplies were reaching Rommel's troops. This would prove decisive in the final defeat of the Afrika Korps. Before he was murdered by Hitler's goons, Rommel wrote in his papers, "The battle is fought and decided by the quartermaster before the shooting begins," and "The bravest men can do nothing without guns, the guns nothing without plenty of ammunition, and neither guns nor ammunition are much use in mobile warfare without vehicles and sufficient petroleum to haul them around."

Fig. 3.6. The German war plan was for Rommel to pass through Egypt, defeating the British, and move north to meet the German army invading Russia at the Baku oil fields on the west side of the Caspian Sea.

Although denied access to Russian and Middle Eastern oil, and with the Rumanian oil fields being bombed by Allied bombers, the Germans still had their synthetic fuel plants supplied by domestic coal and slave labor. By 1943 synthetic fuel production had almost doubled from prewar years and supplied 57 percent of petroleum use and 92 percent of aviation gasoline. Strangely enough, early in the war, Allied Air Command had never thought to target bombers against the synthetic fuel manufacturing industry. In Western Europe, the Allied bombing was initially ineffective until finally in early 1944, General Spaatz asked permission of Eisenhower to bomb the synfuel complexes. Eisenhower gave his permission and by the end of 1944, German fuel production had fallen to one-half of 1 percent of early 1944 production. German airplanes had no fuel with which to fly, and German tanks had no fuel with which to run. The Battle of the Bulge, Hitler's counterattack on the Allies in Europe, failed largely because his tanks ran out of gas and there was no truck fuel for transferring reinforcement troops to the front. German engineers had developed a jet fighter far superior to the prop-driven planes. It was barely used because of lack of fuel. The Germans were out of petroleum.

The Allies were much more successful in maintaining the oil flow to their war machine. Before America entered the war, President Roosevelt initiated the Lend-Lease program to supply the British with oil and tankers. The problem in supplying the British with petroleum was the same as the Japanese had in supplying their home island. The cargo ships were extremely vulnerable to submarines. In 1941 the British were nearly forced to quit the war because of their cargo ship losses to German submarine wolf packs. Also, after declaring war on the U.S., German submarines prowled the East Coast, destroying tankers traveling from the Gulf, around Florida, up the East Coast. Initially the losses were reduced by instituting convoys for greater protection of the cargo ships, a strategy that probably saved England. In a tremendous engineering feat, two pipelines, one to carry crude and the other to carry refined products, were built between the Southwest and the East Coast. The pipelines eliminated the sinking of oil tankers along the East Coast. The U-boat menace, however, continued until the middle of 1943, when the Allies cracked the U-boat code and developed long-range aircraft to protect the transports. In May 1943, the Germans lost 30 percent of their submarines and were never again able to mount an offense.

There is no doubt that U.S. oil fueled the Allied war effort. President Roosevelt appointed Interior Secretary Harold Ickes as the Petroleum Coordinator for National Defense and later as Petroleum Administrator for War. In a rare demonstration of cooperation between government and the various oil companies, Ickes formed a board composed primarily of experienced oil men who oversaw increases in production, transportation, and exploration. During the three and a half years of American involvement in the war, the Allies consumed nearly seven billion barrels of oil. The American oil industry produced, refined, and transported six of those seven billion barrels. It was a supply neither Japan nor Germany could match.

Aftermath of War

Shortly after the war, the center of world oil production would begin to shift from the United States to the Middle East. Before the war began, the Middle East produced only 5 percent of world oil supplies with no increase during the war. In 1943 the very observant Harold Ickes felt the U.S. could not continue with their massive oil production and would soon run low on oil reserves. He thought that shortly after the war, America would become dependent on foreign oil and dispatched a fact-finding team to the Middle East to determine the extent of their oil reserves. The report came back with staggering numbers: a conservative estimate of twenty-five billion barrels of oil, most of it in Saudi Arabia. Ickes felt the U.S. should form alliances with the Middle Eastern countries to insure a supply of oil after the war and form government/business partnerships to produce that oil. The oil companies would have none of it and claimed they had enough oil to fully supply the U.S. for decades after the war. Ickes would be proven correct, and the U.S. would become a net importer of oil by 1949.

Before the war ended, the U.S. and British governments had tried to produce an Anglo-American Petroleum Agreement, which would form a board to monitor world petroleum consumption and suggest production. Again the oil companies would not support the agreement, and it eventually died. Although there was extensive exploration during the war, the only new oil fields to open up in the Americas were around Edmonton, Canada. The U.S. and world companies needed new oil supplies, and all eyes turned toward the Mideast. Ickes left office in 1945 because of disagreements with the Truman Administration. Much of what Ickes had proposed took place, not through any government mandate or business foresight but simply because it had to. Supply and demand forced the agreements between British and American governments, British and American business, and Mideast governments.

By 1948, when all the smoke cleared, through court battles, backroom agreements, and government proposals, western companies and governments had formed alliances with Mideast governments. Socal (Standard of California), Jersey (Standard of New Jersey), Socony-vacuum (Standard of New York, later to become Mobil), and Texaco formed an agreement with Ibu of Saudi of Saudi Arabia, which stated that under the joint company, Aramco (Arabian-American Oil Company), the American companies would produce, refine, and market Saudi oil. Gulf Oil formed an agreement with Kuwait, which allowed access to Kuwait oil, and then Gulf formed an agreement with Royal Dutch/Shell to refine and market the oil. AIOC (Anglo Iranian Oil Company), who had access to Iranian Oil, signed a twenty-year agreement with Jersey and Socony to develop, refine, and market Iranian oil. Aramco completed a pipeline in September of 1950, from Bahrain in Saudi Arabia to Sidon, Lebanon, on the east coast of the Mediterranean Sea. The pipeline (Tapline) would transfer Saudi Oil to Europe and would become a major factor in the recovery of Europe under the Marshall Plan.

Because of the rapid increase in the demand for oil, the U.S. would become a net importer of oil by 1949. The automobile industry was primarily responsible for the increase by almost doubling the number of cars on the road from 1945 to the early fifties. The petroleum industry simply was not prepared for, and could not keep up with, the demand. To cut down on oil consumption and supply energy for domestic water, space heating, and businesses, the natural gas industry was formed. The pipelines built during the war to transfer petroleum from the southwest to the northeast were sold for the transferring of natural gas. Other pipelines were built to link the nation with natural gas reserves.

From the late forties to late sixties, the dependence on Mideast oil became absolute. With new, major discoveries in North Africa (Libya and Algeria), most of the fifties and sixties were a time of an oil surplus. The U.S. Government tried to protect the domestic oil from the influx of low-cost foreign oil, primarily from the Mideast. Initially the Truman and Eisenhower administrations tried voluntary controls which, because of lack of cooperation by the oil companies, failed rather quickly. Finally, in 1959, the Eisenhower Administration, under tremendous pressure from the Texas congressional delegation, instituted protective tariffs and a quota system on foreign oil. In the realm of short-term economics, the tariffs and quotas were a success in protecting American oil companies from the low-cost foreign oil. The act, however, quickly depleted the American surplus and within fourteen years, the tariffs had to be relaxed and quotas eliminated with a much greater dependence on foreign oil.

Also during the fifties and sixties, the international oil companies and oil importer nations were finding that dependence on foreign oil was not always a reliable dependence. To state the problem mildly, the Mideast and North Africa are a geographical areas of social and political unrest. There was a constant battle between the major oil companies and the producer nations over pricing and control. In 1960 the Organization of Oil Producing Countries (OPEC) was formed in reaction to unilateral price cuts by the majors. The five founding members, Saudi Arabia, Iran, Iraq, Kuwait, and Venezuela, exported a total of 80 percent of the world's crude. It was formed along the lines of the Texas Railroad Commission and initially was ineffective due to the world oil glut and the competition for more production and revenue among the members. Once the glut disappeared in the early seventies, OPEC became a very powerful organization.

One factor that affected the reliability of the Mideast source was the large number of wars or coups in the Mideast and North Africa, either when overthrowing one of the governments, fighting among themselves or, quite often, a war about or with Israel. The first Israel-Arab war, in 1956, resulted in the closing of the Suez Canal and shutting off oil shipments by the Mideast producer nations. Because the U.S. still had a large oil surplus and could supply Europe, the act was ineffective. The closing of the Suez and the pipelines, however, did result in the design and manufacture of the supertanker.

In June 1967, the six-day war erupted. Israel, with brilliant military maneuvers, completely eliminated its adversaries in three days. Again, the Mideast producers tried to shut off exports and again, there were enough world resources outside the Mideast and Africa to supply Europe. Again, the oil embargo was ineffective and was lifted in September. It would be the last time the world could supply its oil demands without the Middle East reserves.

By 1972 the conversion of modern civilization to a petroleum world was nearly complete. World oil production had increased from 8.7 million barrels of oil per day in 1948 to 42 million barrels per day in 1972. Every aspect of modern life was changed due to the effect of oil. Oil had replaced coal as the fuel of choice for electrical generation, heating, and transportation. The car became the symbol of the new civilization. People moved to the suburbs and commuted to work in their car. One's identity was determined by the type of car the individual drove and that identity had to be upgraded every few years. A network of superhighways was built to connect parts of the nation with a faster, more efficient transportation system. Soon you could accomplish anything in drive-through, including getting married. Motels replaced hotels as the preferred overnight lodging. Modern materials, primarily plastics derived from oil, became the basis for the new throwaway society. Petroleitus was riding high.

The decade of the seventies brought radical changes to the world petroleum order. U.S. oil production reached its peak in 1970 at 11.3 million barrels per day, and surplus capacity dropped to less than a million per day. In the early months of 1971, the Texas Railroad Commission allowed all out production at 100 percent capacity. There was no more U.S. surplus capacity. Never again would the domestic oil production achieve the 1970s output. The world oil glut had dried up. It was now a seller's market, and prices began to escalate with the producer nations gaining more power. The Environmental Movement came of age and began questioning what effect our modern lifestyles and energy usage was having on the planet.

A major oil discovery (one and a half times the size of East Texas) was made in Alaska in 1967. A pipeline to transfer the oil from Prudhoe Bay to the Port of Valdez in Prince William Sound was planned to be completed by 1972. It was held up for years because of environmental concerns, and the price of oil could not justify the cost of construction. Another large discovery was made in the North Sea, and offshore drilling rigs became a common sight. Offshore oil was discovered in California, but drilling led to a major spill, which produced an environmental backlash that halted drilling and closed derricks.

A New World

A few individuals in the government were becoming increasingly alarmed at the disappearance of American surplus and the dependence on foreign oil (primarily from the Mideast). Yergin, in *The Prize*, discusses James Placke, a Mideast diplomat who was connected to the oil industry in the Middle East for ten years. In late 1970, Placke wrote a very thorough commentary on the status of Mideast oil and politics. He felt the whole balance of power would soon change when the producing nations overcame their differences and worked in unison to raise prices. Also, those in the Mideast who would use oil as a weapon were gaining influence and oil would soon be a bargaining tool. The U.S. ambassador felt the commentary was so insightful, he added his name to the report and sent it on to Washington. Nothing more was heard of it.

With the rapid rise in prices in the early seventies, oil supplies were becoming a political hotseat. George Shultz, Nixon's secretary of state, formed a committee to investigate the tariff and quota system. Their report included a recommendation to eliminate all quotas and replace them with a simple tariff. The oil industry and their politicians rebelled, and the report was shelved. It would take more drastic events to eliminate the quotas.

James Adkins of the State Department wrote a report stating the era of cheap energy was over and the U.S. should take steps to "reduce consumption, raise domestic production" and look for more "secure sources." In a secret report to the president, he recommended increased coal use, development of synthetic fuels, increased conservation efforts, and research to get beyond hydrocarbons. The report was dismissed, with Nixon's aide, Ehrlichman, stating, "Conservation is not a Republican ethic," a sentiment still echoed thirty years later by Vice President Cheney in 2001. Adkins later published an article in the *Foreign Affairs* magazine, which summarized his thoughts on U.S. oil problems and listed the steps that should be taken. It appears the only individuals who read and took note of the article were the Middle East oil producers.

On October 6, 1973, the third Arab-Israeli war, the Yom Kippur War, started with a surprise attack on Israel by Egypt and Syria. It would forever change the face of the oil industry and America's feeling of security toward Mideast oil supplies. The whole post World War scenario would be turned upside down. Anwar Saddat, the New Egyptian leader following Nassar, planned the war because he felt Israel and the United States were refusing to negotiate following the losses the Arabs incurred after the six-day war.

Israel and the United States were caught completely by surprise in what has been called Israel's Pearl Harbor.

Initially the Egyptians and Syrians made rapid advances and were on the verge of eliminating Israel. Ultimately, with major U.S. supplies, the Israelis made a series of counter-offenses and drove the Egyptians and Syrians back to their original lines, in the process nearly destroying the Egyptian army. The problem, as the Arab Nations saw it, was the way in which the United States resupplied Israel. The original plans were to fly in the supplies at night in a subtle manner. Because of strong winds at the refueling station, the C5As were forced to land in Israel in the middle of the day, showing very obvious and blatant support for Israel. This angered the Arab oil-producing nations, who were at the same time having difficulties in negotiating with the western companies over their share of oil profits. The result was the oil embargo. For the first time, the producer nations were able to use oil as an effective weapon.

The effect on the U.S. and world economies was devastating. In September 1973, OPEC raised the posted price of oil by 70 percent, to $5.11 a barrel and again in December, to $11.65 a barrel. To punish the friends of Zionism, the Arab members of OPEC (OAPEC) also agreed to an embargo, which would cut their oil production by 5 percent and continue cutting by 5 percent each month until their goals had been realized. There was no policy in place by the consumer nations that would help eliminate, replace, or share the oil shortfalls. The burden of allocation then fell to the major oil companies. In retrospect, they seemed to have accomplished about as fair an allocation system as was possible at the time. Several countries raised strenuous objections and felt they should be supplied fully, primarily Britain, Japan, and France. The majors held fast, and under a policy of "equal suffering," continued to allocate oil in what they felt was a fair manner.

During the previous twenty years, a time of the petroleum excess, the U.S. economy had flourished. With the oil embargo and price increases, the world industrial nations plummeted into a recession. The U.S. GNP fell 6 percent between 1973 and 1975, and unemployment rose to almost 10 percent. The shortage of gasoline forced long gas lines at the pumps with many gas stations closing every other day. Prices skyrocketed. The embargo was called off on March 18, 1974, but the world of the petroleitus would be forever changed. In the following years, OPEC would determine the state of the world economy.

Also, in retrospect, it is difficult to understand why the industrial nations were so unprepared for the oil shortfall and price increases. In February 1974, the United States convened an energy summit among the industrial nations to discuss the world energy problem, almost like closing the barn doors after the horses get out. The conference did result in an emergency sharing program if more shortfalls occurred and the formation of the International Energy Agency, which would help manage the energy programs of the western nations. The lack of foresight on the part of the United States can be blamed in part on the Watergate Incident. President Nixon was totally preoccupied with Watergate and gave very little time or importance to Middle East affairs, so much so that at the height of the Israeli-Arab war, when the super powers were squaring off against each other with their nuclear arsenals, Nixon's advisors did not even bother to wake him during a crucial meeting. The mantle of leadership fell to Henry Kissinger who, previous to the war, knew next to nothing about petroleum.

In the next few years, OPEC began to flex its muscles. By the mid-1970s, all concessions were taken over by the producer nations. All the productions facilities in the OPEC countries were nationalized. The producer countries would now set production quotas and price. The former petroleum production companies (ARAMCO, AIOC, Gulf-Royal Dutch partnership, and British Petroleum) were reduced to buying and selling oil on the market. During this time, the Saudis (being the world's largest producer) became a moderating influence in OPEC. The Shah of Iran (world's second-largest producer) would constantly strive for higher prices.

The U.S. struggled to formulate an energy policy. Nixon set price controls and tried to return to pre-embargo prices, which stimulated strong economic growth. Since the U.S. no longer had control over oil prices, this was impossible. One could write several books, dissertations, and lectures on the mess that was brought about by price controls. Basically it was a disaster. By late summer 1974, due to Watergate, Nixon was forced to resign. The economy was in shambles. After Vice President Ford took over for Nixon, he proposed a ten-year plan to build two-hundred new nuclear power plants, develop two-hundred fifty new coalmines, and build one-hundred fifty new coal-fired electric power plants, thirty major oil refineries, and twenty major synthetic fuel plants. Also immediately after the oil embargo, Congress gave the green light to the Alaskan Oil Pipeline and passed a law mandating fuel-efficient standards of the automobile industry. Over a ten-year period, the fuel efficiency of the automobile would be required to double, from 13 mpg to 27.5 mpg. At that time, one out of every seven barrels of oil used in the world was consumed on U.S. highways. Of all the proposals and legislation made during the Nixon and Ford administrations, only the construction of coal-fired electric power plants, mandated gas mileage for automobiles, and the Alaskan Pipeline had any effect on the U.S. energy situation.

The Struggle for an Energy Policy

In 1977 Jimmy Carter became president while promising to bring a new energy policy to the country within the first ninety days of his administration. To complete that task, he made James Schlesinger the energy czar and formed the new Department of Energy. Schlesinger was a brilliant Ph.D. economist who had formally worked in the Rand Corporation before joining the Nixon and Ford administrations. He proposed the very controversial idea that conservation should be central to any energy policy and stated that implementing that policy should be the "moral equivalent of war." Carter later adopted the phrase and used it often in his talks.

Schlesinger, by emphasizing conservation, coal-fired electric power plants, and development of renewable energy sources, felt they could lift the price controls imposed by the Nixon Administration. Petroleitus, led by special-interest groups, was having none of it. They wanted, and felt they deserved, their cheap energy to supplement the affluent American lifestyle. To Schlesinger the problem and solution were very clear. The U.S. had to separate economic growth from the price of petroleum and natural gas. There were relatively large amounts of world oil supplies remaining; however, the time of cheap oil, which had fueled western economies for the previous fifty years, had ended. Americans had to start looking for other sources of energy. The most expedient way to accomplish this task was to let oil and natural gas prices rise while funding research and development of alternatives.

Carter and Schlesinger got a lesson in the American democracy. Citizens and interest groups do not usually consider long-term problems but are more concerned over immediate needs. Interest groups had long discovered how to employ political power. Schlesinger equated his time spent at committee hearings on fuel price controls with time spent in hell. Finally, however, the National Energy Act was passed. The act was a very watered-down version of what the Carter Administration was trying to pass, but it did allow some increase in energy pricing and supported research into alternatives.

In the following years, the ending of the oil embargo and increase in oil prices led to renewed economic growth and extensive world oil exploration. Because of the nationalization of all Mideast oil sources, most of the exploration occurred in North America or the North Sea. Before the new discoveries could be integrated into the world market, the Mideast had other surprises for Petroleitus.

Exit the Shah

In the early years of OPEC, the Shah of Iran had been an extremist in his demands for increasing petroleum prices. Since the 1973 war, the Shah had changed hats and along with the Saudis had become a moderating voice in OPEC. Also, he had become a darling of American foreign policy as a buffer between the Mideast and the Soviet Union. His government tried to introduce progressive western values into Iran but was violently opposed by Islamic fundamentalists. By late 1978, the Shah was in serious trouble with the Iranian populace, who felt he was not really one of them and was a lackey of the west.

The Carter Administration had no clear policy as to support the Shah or let events play out and try to find a successor with whom they could work. As a result of the "no policy" policy, the Shah received vacillating remarks from the U.S. Government and did nothing. One U.S. diplomat later remarked that we "would have been better off flipping a coin" and going with that policy. In January 1979, after the oil fields had completely shut down, the Shah left Iran for the last time.

The Iranian debacle demonstrates a complete misunderstanding of the Iranian people and Islamic fundamentalism by both the Shah and the American Government. In the months leading up to his overthrow, not only did the American Government have a "no policy" policy but so too did the Shah. William Shawcross in *The Shah's Last Ride* describes the conflict within the Shah's government between his advisors who wanted to imprison those were trying to overthrow his government and his other advisors, who felt he could work with them. As a result of this conflict and the lack of support from the American Government, the Shah also did nothing. Here we see the beginning of the misunderstanding of Islamic fundamentalism that has plagued our Mideast policy to this day.

As the second-largest petroleum producer in the world, the closing of the Iranian oil fields sent a second shockwave through the world petroleum industry. Much of the fear was over the possible spreading of Iran's Islamic revolution and further interrupting the flow of Mideast oil. Islamic fundamentalism combined with nationalism and rejection of western values was becoming a powerful force in many Islamic states. America as a symbol of western values would become the "great Satan." When all was said and done, however, it was more a perceived shortage than much of an actual shortage. The Saudis increased production, which left only about a 4 percent difference between world supply and demand. But, the effect was devastating.

The panic had begun and would carry on a life of its own. Japan's oil supplies and British Petroleum were hit the hardest because of their dependence on Iranian oil. The Spot Market became the place to find oil. Long-term contracts between sellers and buyers were broken, with the seller's oil quickly appearing on the spot market where the buyer could purchase it at a 20 percent to 50 percent price increase. The Saudis once again were the moderating influence and kept the panic from spreading even further. They held prices and forced all of their buyers to hold prices if the oil was resold. Iranian oil soon came back on the world market but at a much lower volume. When Iran began producing oil again, the Saudis returned to their pre-1978 production level, which did not make up for the lower volume of Iranian oil. These shortages and perceived shortages again led to a destabilizing effect on the world economy and a major world recession. Gas lines and panic buying at the pumps signaled the end of Jimmy Carter's Administration.

Other events of the time seemed to conspire against the Carter Administration. In March of '79, the Three-Mile Island incident occurred, which to many spelled the end of the nuclear electric generation industry. He fired two of his cabinet members and three others, including Schlesinger, resigned. Carter tried to bolster his image by announcing a synfuel project that would supply two and one-half million barrels a day. Nothing came of it, although with the world recession and energy demands decreasing, the major companies continued to build up their inventories at extremely high prices. No one seemed to know what was going on and by late 1979, the oil industry was in complete chaos. The final straw was the attack on the U.S. Embassy in Tehran and the taking of the American hostages. Carter was now perceived by most of the American populace as a completely inept president.

As a result of the panic buying by the majors, OPEC (except for Saudi Arabia) exercised their profit motive and continued to increase prices. There was no reason to believe they would not get them. Officially, in late 1979, the OPEC rates were thirty-two dollars a barrel, but some companies were paying as high as fifty dollars on the spot market. The Saudis tried to warn OPEC of the excesses of greed and felt the increases would ultimately be back to haunt them. The more radical members of OPEC were not buying it and felt God would keep the demand high. Oil gluts, however, were signaling the beginning of the end of OPEC's power.

On September 22, 1980, in a complete surprise, Iraqi warplanes and troops attacked Iran. Now OPEC members were fighting among themselves. In an extremely bloody, suicidal war, Iraq lost most of its oil production and Iran soon lost all of theirs. The loss of Iranian and Iraqi oil removed four million barrels a day from the world market. Spot prices jumped to forty-two dollars a barrel. Many thought this would send a third shock through the world economy and perhaps spell the doom of western civilization.

OPEC met in December 1980 and decided to raise prices to thirty-six dollars a barrel. This time, however, no one was buying. The western powers, including Japan, were finally agreeing on a unified policy, and the world oil scenario was becoming clearer. With conservation measures taking effect, a soft economy, increased production by the Saudis, increased production in Mexico, Alaska, and the North Sea, the western powers found they no longer needed the rest of OPEC oil. The increased production more than made up for the losses in Iran and Iraq. A world oil glut had arrived. The U.S. and world economies were again able to pick up speed. OPEC could no longer dictate the world economic situation.

Good Times for All

Petroleitus was back in the saddle again and would be riding high for the next twenty years. Ronald Reagan was elected president and would ride the oil glut through a period of economic growth. He welcomed in a time of optimism and prosperity in sharp contrast to the preceding fifteen years. Conservation measures put in place during the Carter Administration, including government mandating of fleet gas mileage for the automobile industry, converting from oil to coal-fired electric generating plants, and more efficient appliances in the home and industry, led to a massive drop in energy consumption. The petroleum share of the energy market dropped from 53 percent in 1978 to 43 percent seven years later.

The only industry hurting in the new economy was the oil companies and the oil-producing countries. The price of oil continued to fall, and OPEC became a cartel to set production quotas for controlling price. The quotas depended on the integrity of member nations and most, except Saudi Arabia, violated their quotas by selling on the spot market. The value of world oil companies fell drastically and soon became lower than the value of oil reserves they owned. Many oil companies were bought out by other larger or smaller oil companies. Texaco purchased Getty. Occidental purchased Cities Service, and Chevron bought out Gulf. British Petroleum gained control of Sohio and later purchased the whole company. A few companies bought out themselves just to raise their value. With the rapid fluctuations in oil prices, investors began trading in petroleum futures. By 1985 the supply exceeded demand by 20 percent, and world leaders no longer focused on petroleum supply as a problem.

In late 1985, oil prices collapsed, from a high of thirty-two dollars a barrel to a low of ten dollars. On spot markets, it sometimes sold as low as six dollars. American and European producers demanded protectionist tariffs from foreign oil. A panic went through the industry. Finally, in the summer of 1986, OPEC and non-OPEC producer countries, along with industry spokespeople and government representatives, met to prevent a collapse. A quota system and a price of around eighteen dollars a barrel was agreed on. The price was meant to be high enough to prevent the implementation of tariffs and low enough to prevent research into, and implementation of, alternative fuels.

The late eighties found a greater cooperation between most of OPEC countries and the West. As a result of the Iraqi-Iranian war, the U.S. found itself as the protector of Mideast oil supplies. Middle-Eastern ships would be reflaged under the American flag, which allowed the U.S. Navy to escort them. Some of the Mideast countries formed joint ventures with or purchased shares of American and European petroleum companies. This would guarantee access to western refineries and gas stations.

In 1988 the United States elected a new president, who was very familiar with the oil industry. Before going into politics, George Bush had spent his younger years chasing around the United States developing a business in the oil industry. He was very aware of the relationships between oil and politics and had spent some time in the Middle East. His knowledge of the Middle East would be tested in the coming months of his presidency.

In 1987 the Iraqi-Iranian war ended with no real winner but with Saddam Hussein claiming victory. He then threatened the West with the oil weapon, but nobody cared. The world no longer needed Iraqi oil. In 1990 Hussein invaded Kuwait to gain more control over Mideastern oil resources. The oil weapon was turned against him with a boycott of Iraqi and Kuwaiti oil. Under Bush's leadership in forming a coalition, the U.S., European, and Middle Eastern powers sent a military force to drive Hussein out of Kuwait. The Iraqis were easily defeated, but on leaving Kuwait, they started fires in virtually all the Kuwaiti oil wells. Within months the fires were extinguished, and the Middle-Eastern war produced only a mild recession in the world economy. The mild recession may have cost Bush a second term. Although defeated, Hussein would not disappear from the world scene and, for the next ten years, would be a thorn in the side of western and Middle-Eastern powers.

In 1989 the Cold War ended with the breakup of the Soviet Union and the end of the Iron Curtain. With Russia and the former Soviet states converting to capitalist economies, more oil was released onto the world market. Russian companies formed joint ventures with America companies to gain access to the American market. Over the past ten years, Russian oil has gained greater influence in the world market, and now Russia has a major say in the regulating of world oil prices.

In the late eighties, with world economies strong, the environmental movement gained strength. In 1986 the Chernobyl meltdown occurred with radioactive gases being released over much of Europe. The accident signaled the end of the nuclear industry over much of the world. The year 1989 brought the Exxon Valdez spill in pristine Alaskan waters. The cleanup cost was nearly two billion dollars and focused the environmental movement onto the petroleum industry. Evidence was mounting that showed an increase in global temperatures brought on by the combustion of carbon-based fossil fuels. Although the Reagan Administration (which was certainly not environmentally friendly) dramatically reduced the funding of alternative energy fuels in the U.S., Japan and Europe have continued increasing their research into alternative fuels and conservation. They now spend much more than the U.S. in those areas.

Bill Clinton was elected president in 1992 and for the next eight years, the American economy, with low, stable oil prices, showed the largest growth in forty years. Petroleitus was riding high but demonstrating a very short memory. Many of the conservation policies put into effect by

Schlesinger and the Carter Administration were either thrown out, ignored, or bypassed. Nothing symbolizes this more than the SUV (the Sports Utility Vehicle). In the seventies, Congress had mandated that over a ten-year period, each U.S. car manufacturer would be required to double their average fleet gas mileage and lower emissions. This policy, along with the construction of coal-fired generating plants, were the two policies that made a major difference in our oil consumption and allowed for the economic recovery from the seventies. With gas prices low and a very short memory, however, manufacturers and consumers could now return to the large gas guzzlers of the sixties and seventies. The government had passed a law that would prohibit the manufacture of cars with such low gas mileage. To bypass the law, SUV and minivans could be classified as "light trucks," which allowed for lower gas mileage and higher emissions standards. Since the late eighties, the number of light trucks on American highways has increased dramatically and in 1998, one of every two vehicles sold in the U.S. was a light truck. Also since the mid-eighties, US and world petroleum consumption was again on the rise.

A major factor in the rise of world consumption has been the rapid industrialization of many of the Asian countries, "the so-called Asian Miracle." The most dramatic rise in petroleum consumption has occurred in the Asian Pacific region with the area now using 27 percent of the world oil resources and increases of 2.6 percent per year. It is more than the U.S. at 26 percent of world oil consumption and a .1 percent rise per year, or Europe at 22 percent of world consumption and a .2 percent rise. China is now the number two consumer of Mideast oil.

The world is desperately searching for more petroleum reserves. It appears, however, that most of the major discoveries of the easily assessable oil are over. In 1995 U.S. Senate hearings were held on the opening of the Arctic National Wildlife Refuge (ANWR) to petroleum exploration. It is the most promising new field in the U.S. and most probably the world. It is located in the coastal plains of the refuge, a place designated as ANWR 1002. Best estimates of economically recoverable oil from that field, by a multitude of state and federal agencies and private companies, are about a 5 percent chance of retrieving four billion barrels of oil. This would supply the voracious American appetite for petroleum for about two-hundred days.

Where Are We Going?
The Extinction of Petroleitus

One has to be very wary when trying to predict the future. If because of the difficulties in understanding the many variables that determine our future and the very real chance of being wrong, we neglect to make an attempt that may lead us to wiser decisions, we would be failing future generations. So with the information at hand, let us attempt to construct a vision of the next twenty-five years, and as new information becomes available, we will modify that vision. If we are to error, let us error on the side of safety.

In the 2000s, 70 percent of U.S. petroleum was imported. Canada is our largest supplier, followed by Mexico, Venezuela, and Saudi Arabia, in that order. The Alaskan oil fields (North Slope) produce about 25 percent of U.S. domestic oil and declining, with the lower forty-eight states, primarily Texas and Oklahoma producing the rest, and declining even more rapidly. Between 2008 and 2014, the Alaskan Pipeline will be closed if no new fields are developed (this includes ANWR 1002). When finished, the total amount of oil retrieved from the North Slope fields will be approximately twelve billion barrels. Best estimates of ANWR 1002 would place it at about a third the size of the North Slope. If these estimates are close to being correct and our oil consumption remains constant, by 2012 the U.S. will only supply about 8 percent of its own petroleum needs. We will then be completely dependent on world resources.

Over the past few decades, there have been primarily two methods for predicting the amount of oil remaining and when we will arrive at peak production. The first has been to try to determine the amount of oil available in all remaining reserves, both discovered and undiscovered, and then to determine how fast we are using it. By dividing the two numbers, we should have the time left for petroleum species. Determining the amount of reserves remaining is an inexact science at best.

The second method is a statistical method developed by M. King Hubbert that looks at oil consumption production on a larger overall scale, the so-called Hubbert's Peak Method. In 1969 Hubbert made his first prediction that we would arrive at peak production around 2000. In 2005, based on the latest information and using Hubbert's methods, Kenneth Deffeyes has predicted that peak oil production occurred late in the year 2005. Both Hubbert and Deffeyes arrive at total world overall oil production (total world allotment of oil) as around two trillion barrels, of which about half has been used. We will look at Hubbert/Deffeyes method in Appendix A, along with the view of some coracopians.

In 2000 the world energy council of the United Nations Statistical Division did a survey of proven world energy reserves. That is how much oil is left. They arrived at a figure of one hundred twenty-one billion metric tons of oil, which translates to almost nine hundred billion barrels. This is quite close to Hubbert/Deffeyes calculations. Of the nine hundred billion barrels, 30 percent of the reserves are found in Saudi Arabia and 63 percent are found in all Middle-Eastern countries.

Also in 2000, the U.S. Geological Survey (USGS) invested two hundred person years into completing an assessment of world oil reserves. They found the total world allotment to be approximately three trillion barrels, of which approximately one trillion have been used. At present consumption and growth rate, the extra one trillion barrels found by the USGS would only add about ten to fifteen years to the supply before the well runs dry.

This would seem to be at least some time to burn more oil and gradually shift to alternative fuels, but the problem arises with supply and demand. The recession of the seventies, which completely threw western economies into turmoil, was a result of only a few percent differences between the supply of oil and demand for oil. In March 1998, an article was printed in *Scientific American* which, through an analysis of world oil sources and the present demand and growth rate, showed the demand becoming greater than the available supply occurring sometime in 2007. This may be off by a few years, and there will still be large amounts of petroleum left in the earth, but the time of even moderately cheap oil will be over. The effects on our economy will be devastating.

One problem with the USGS survey is that it relied on the honest reporting of the various oil producing countries and companies. It is an advantage for the countries and companies to overstate their reserves to give their companies an appearance of greater worth. Since the initial report, several companies have admitted they over-stated their reserves by as much as 30 percent. It now appears that at the time of completing the writing of this book (2007), virtually every oil-producing company or country in the world is pumping oil nearly full bore. There are very little reserves remaining.

Also, since most of the available oil is found in the Mideast, if something unforeseen were to happen in that area that closes the oil spigot, western economies would collapse. The events of September 11, 2001, emphasized the instability of that area. Most of the hijackers of the planes that destroyed the World Trade Center and damaged the Pentagon were from Saudi Arabia. Most of the members of the Taliban and Bin Laden, who sponsored the attack, are from the Mideast, primarily Saudi Arabia. If the Islamic fundamentalist terrorists gain control in any of the Mideast oil-producing countries, our supply of oil will be drastically reduced.

In 2000 George Bush's son, George Bush Jr., was elected president. In reaction to the September 11 bombing, a coalition led by the United States invaded Afghanistan to eliminate the terrorists responsible. The invasion was only partially successful, and the terrorists threats remain. In 2003 the United States and Britain invaded Iraq to eliminate Saddam Hussein. Hussein was eliminated but at this time, it appears that unless there is a radical change in the atmosphere created in Iraq by the invasion, the world will certainly be no safer. The invasion seems to have been precipitated on the same misunderstanding of Islamic fundamentalism by American Foreign Policy that began with the ouster of the Shah. The hatred of the United States by Islamic fundamentalists has only increased along with the possibility of more violence in the area.

China and India are also becoming major players in the world energy scenario and are requiring more petroleum. The tremendous growth of the economies in both countries, around 10 percent per year, are even more tied into petroleum consumption than we are in the U.S. This is causing even more strain on our supplies of world oil and are drastically increasing prices.

In the 1995 Senate hearings on Alaskan oil reserves, Senator Dale Bumpers from Arkansas made an insightful statement, to which nobody seems to be listening.

> "I want to make the point that ANWR will be produced, the 1002 lands will be produced, if there is, in fact, oil there. Nobody really knows for sure whether it is there, whether it is economically feasible to produce it, but I come down on the side that there is absolutely no need to do it now.
>
> It would raise the world's reserves of oil by four-tenths of one percent under the most, what should I say, optimistic estimates. It would provide the United States with 200 days of oil.
>
> Now, when you consider some of the environmental considerations of what can and might happen in that area, I am not sure all of that is worth it.
>
> And finally, Mr. Chairman, I am not suggesting we should not find oil, and that we should not encourage the oil companies to produce more oil and find it, but I want to say we are facing a train wreck, energy wise, not 10 years from now, probably not 20 years from now, but it is coming.
>
> Everybody knows that this country is so far from becoming energy independent, it is not even debatable.
>
> I just think there is a psychological factor involved that as long as we can get oil from Saudi Arabia and the Middle East, Prudhoe Bay, and maybe the 1002 area of the Arctic Wildlife Refuge, we are not going to become the conservation society we are going to have to become.
>
> Everybody knows it, and yet we keep procrastinating and putting it off, and assuming, and trying to mislead people into believing that somehow or other it is just out there, and if you just turn us loose and let us find it, this country will become energy independent.
>
> Our energy independence is not ever going to be based on oil. It is going to be based on *conservation and alternative energies*." (Author's italics).

Chapter Four

*The Rise of the Utilities
Natural Gas and Electricity*

Natural Gas

Like petroleum, the ancients were aware of natural gas and noticed that lightning could ignite the gas seepage and create a burning spring. The most famous of these burning springs was on Mount Parnassus in Northern Greece. Around 1000 B.C. a goat herdsman noticed the fire and a temple was built around it. A local priestess, who became known as the Oracle at Delphi, expounded prophesies for those who visited the site. Although the Greek Oracle was the most famous, natural gas springs were also used as religious sites in Persian and India. The Chinese were the first to recognize the heat value of natural gas around 500 B.C. They used bamboo pipes to transmit the gas to where they could burn it to evaporate sea water and separate the salt.

An early form of natural gas became available in Europe during the later part of the eighteenth century. The gas, manufactured synthetically, was primarily carbon monoxide, hydrogen, and sometimes methane, and became knows as manufactured gas. It was produced through a variety of processes, which usually included passing hot steam over heated coal. At times other hydrocarbons, including animal fats, were used instead of coal. By 1780 most London streets were illuminated using manufactured gas. The technology of producing manufactured gas was transferred to the United States around 1816, when Baltimore, Maryland, installed gas streetlights.

There was an abundant amount of natural gas available in the U.S., and sites from New York to California were recorded by early explorers and missionaries. Native Americans ignited seepage from these sites and at times used the heat and light. One major source of the bubbling gas was near the eastern tip of Lake Erie in New York and is considered the birthplace of the natural gas industry in the United Sates. In 1821 William Hart, sometimes considered the father of the natural gas industry, dug the first gas well outside of Fredonia, New York, and used the gas as illumination in local houses. Soon area businessmen expanded on his work and formed the Fredonia Gas Light Company, the first gas company in the U.S. and the beginning of the utility industry. Other areas started gas exploration, and by 1900, natural gas had been discovered in seventeen states. When gas was discovered near larger cities, it began to compete with manufactured gas as a source of illumination.

From the beginning of oil exploration, gas was found to be a byproduct of the petroleum industry. When Drake drilled the first oil well in 1859, it was determined that natural gas supplied the pressure that forced petroleum up the pipe. Originally the gas was vented and/or burned. In 1872 a wrought-iron pipeline two inches in diameter was built to carry the gas a distance of five miles from Drake's field to a nearby town, the first successful metal gas pipeline. Scientists soon found three types of natural gas occur naturally in nature. The first is associated gas that is found with petroleum and originally supplied 100 percent of the U.S. demand. As oil exploration continued, dry wells were drilled that contained only natural gas, no petroleum. This non-associated gas now supplies about 75 percent of the nation's needs. Later, as deeper wells were drilled under higher pressures, a cross between a liquid and gas called gas condensate was discovered. This condensate can be refined into various products, including natural gas, butane, and propane.

In 1885 the Wellsbach mantel lamp was developed. When gas combustion occurred in the presence of the mantel, it burned with an extremely bright, incandescent glow. This increase in the efficiency of lighting made manufactured or natural gas the flame of choice in most large cities. It would be short lived, however; with Thomas Edison flipping the first switch on an electric generator in 1882, electric lights rapidly replaced the gas light. Just as in the case of petroleum, new technological developments would lead to new markets for natural gas as an energy source. The first of these was the Bunsen burner, which was developed in the middle of the nineteenth century. The Bunsen burner mixed air with the gas before it was combusted. This air/gas mixture gave the much hotter blue flame we now associate with natural gas and led to its use in cooking stoves and heating.

The primary technological advancements that would help natural gas develop new markets was the improvements made in pipeline and pump technology. Natural gas had a market if it could be transferred to the population centers that made up the market. The first pipelines were wood, and later wrought iron was used. The first long-distance pipeline, built in 1881, was cast iron and transferred gas one-hundred twenty miles from the fields in Indiana to Chicago. The gas came out of the ground under high enough pressure (five hundred twenty-five pounds per square inch) so the inefficient pumps available at the time were not required. The pipes were bolted or riveted together using couplings. In 1903 an eighteen-inch pipeline was built from the Ohio River to Cleveland, a distance of one-hundred eighteen miles. In 1909 a twenty-inch pipeline was built that extended from West Virginia across Kentucky and the Allegheny Mountains, then west and north to Lake Erie.

The early pipes were very crude by today's standards, and if even a small leak occurred with the gas under pressure, the results could be catastrophic. Thicker walls did not help the strength of the pipes because the weakness was in the seams and couplings. In the early 1900s, steel pipe began to replace cast iron and in 1911, oxy-acetylene welding was developed. This dramatically improved the manufacture and coupling of pipe. In 1922 electric welding was developed. Everything was in place for long-distance gas pipeline industry.

As the technology improved, the industry began to build larger diameter pipes, and the gas was transmitted under greater pressures. The large pipelines increased the economies of scale, and soon natural gas completely replaced manufactured gas. By the late 1920s, over a dozen transmission pipelines were built. The lines, all greater than two-hundred miles in length, carried gas from the Monroe field in Louisiana, the Panhandle-Hugoton fields, which spread three-hundred miles across northern Texas, Oklahoma, and southwestern Kansas, and the San Joaquin field in California into population centers in their areas.

As the longer pipelines were constructed, it was found that quite often, during times of peak winter usage, the supply of gas from a well could not keep pace with the demand. The pressure in the reservoirs and pumps that were used to supply the gas could not produce the volume needed. Suppliers came to the conclusion they needed some method to store large volumes of natural gas near their end usage market. Initially they tried large tanks similar to the ones used to store oil or water but had very limited capacity to store gas. This limited storage capacity and lack of pressure to adequately supply natural gas during times of peak demands constrained the development of the industry. Finally someone hit upon the simple but brilliant idea that since natural gas came from the ground, why not pump the gas back into the ground and store it in large underground caverns? The first of these, the Zoar Storage Facility, was built in New York in 1916. Next was the Menifee Reservoir in Kentucky in 1919 and Queen Reservoir in Pennsylvania in 1920. As these huge underground storage facilities were added, suppliers could pump gas into them during times of low demand and withdraw the gas at times of peak demand. The storage process also allowed for smaller pipelines to be built from the well to the storage facility, since it did not have to supply the peak winter demands.

Also with the production of natural gas, producers had found there were other hydrocarbons that could be separated and sold as independent products. Some of these include propane, which developed into a fuel industry that would become the primary fuel for rural America before REA (Rural Electrification Act). Ethane was used to produce ethylene, an extremely important petrochemical. Iso-butane was used in high-octane fuel and supplied much of the aviation gasoline in World War II. Other trace components of natural gas include heavier hydrocarbons (butane, pentane, hexane, and heptanes), which are used in motor fuel production. Carbon black was produced from natural gas, which is used in the production of ammonia and some medical supplies and also became a major market during Word War II.

Government Regulation

Initially, all natural gas pipelines and sales were intrastate and therefore governed by the states in which they operated. With the growth of long-distance pipelines, the natural gas industry became interstate and was therefore under the governance of the federal government. West Virginia was an early supplier of natural gas for the northeastern United States, which brought quite a lot of income to the state. The state legislature was concerned about sufficient supplies for the citizens of West Virginia and mandated they be given priority treatment. This did not sit well with the gas consumers in Pennsylvania and Ohio, who took the case before the Supreme Court. The court decided the mandate would interfere with interstate commerce, and the legislation was ruled unconstitutional.

Just as the petroleum industry was running into problems in the late twenties and early thirties, which brought federal investigations, so too was the natural gas industry. In the late twenties, Rockefeller and Morgan interests controlled 36 percent of the natural gas production and almost 90 percent of natural gas transportation. Some of their same holding companies that developed manufactured gas from coal also had vested interests in the natural gas industry and had formed natural gas companies. The holding company would not allow natural gas pipelines into an area that could compete with their manufactured gas until the manufactured gas equipment was completely depreciated or completely worn out. At times, a pipeline of sufficient capacity would pass within a few miles of an urban center and still not be connected to the city or allowed to service it. In some areas, consumers were being charged more than twice the amount for manufactured gas as the natural gas they were not allowed to access. Also, the price for natural gas varied tremendously from state to state and sometimes even from city to city within a state. As a result of the inequities, a Cities Alliance, representing one hundred midwestern urban areas, was formed. The Cities Alliance had no political or economic ties. Their only agenda was to bring fairer pricing and access to pipelines for their urban areas.

Because of the holding company's monopolistic control of the energy industries, federal investigations began in 1928. In the early 1930s, the Federal Trade Commission (FTC) began an investigation into the natural gas industry. When FDR was elected in 1932, the investigations were continued and many of the old boys from the natural gas industry who were on the investigative boards were dismissed or fired. Thomas Corcoran was put in charge of the hearings and declared a personal crusade against the power of holding companies. Because of the previous investigation of standard oil, many Americans felt holding companies were a threat to the democratic and free market characteristic of American society.

In their investigations, the FTC identified sixteen specific evils of the monopolistic structure of the natural gas industry. Eight of these were specific to the unfair pricing and availability of natural gas. The eight were 1: excessive cost of natural gas because of excessive competition in drilling wells; 2: costly struggles between rival natural gas companies to conquer or defend territories of distribution; 3: excessive and inequitable variations in city gate rates for natural gas among different localities; 4: excessive profits in many natural gas sales between affiliated companies; 5: exploitation of subsidiaries of natural gas companies through fees for construction, management, and promotion; 6: exaction of excessive bonuses or commissions by investment bankers in connection with financial transactions with natural gas companies in certain instances; 7: exaction of excessive bonuses or commissions by officials of certain companies in connection with sales and construction of properties; and 8: misrepresentation of financial conditions, investment, and earnings of some natural gas operating and holding companies. (The final five sound like the Enron debacle.) The FTC, in their findings, recommended greater federal control of the natural gas industry.

The study was used as support for the passage of the Securities Act of 1933. It also led to the enlargement of the jurisdiction and functions of the Federal Power Commission (FPC) with the Federal Power Act of 1935. The act also created the Securities and Exchange Commission. From 1935 to 1937, hearings continued on the natural gas industry with testimony largely supplied by industry spokespeople and representatives from urban communities, primarily by the Cities Alliance. The hearings resulted in the Natural Gas Act of 1938, which gave the FPC control over the price charged for natural gas by interstate pipelines. It also encouraged the establishment of just and reasonable rates without undo price discrimination among customer classes. The FPC also assumed control over the extension, addition, and discard of pipelines.

As a result of the Natural Gas Act of 1938, the structure of the industry would not change for the next forty years. Basically, all gas producers would find and develop the gas fields. The producers would then sell the gas to pipelines companies, who would transport the gas to various parts of the country. The pipeline companies would sell the gas to local distribution companies, usually utilities, who would then sell it to the end user. The price of gas the producers charged the interstate pipelines was regulated by the federal government, as was the price the pipelines charged distribution companies. The price the distribution companies charged the end user was regulated by state or local government. Although the expansion of the industry was slowed down by the Great Depression, by the end of the 1930s, natural gas pipelines crisscrossed much of the south central, east central, and the Eastern Great Lakes area of the United States. By 1940 parts of thirty-four states had some natural gas services. In most of the states, however, thirty-one of the thirty-four, only a small percent of the households had access to natural gas. By December 1940, everything, with the exception of large efficient pumps, was in place for the expansion of the industry into the rest of the country. This expansion would have to wait until the late forties and fifties, as World War II got in the way.

Because of the shortage of steel and labor, there was very little building of pipelines during the war. The one exception was the federal government granting the Tennessee Gas Transmission company permission to build a gas pipeline from the Gulf Coast to the Appalachians. Also, as was mentioned in chapter three, to avoid the sinking of oil transport ships by German submarines, in a Herculean effort, the government had two pipelines built to carry petroleum products from the southeastern United States to the eastern seaboard. After the war, these pipelines were converted to transporting natural gas and sold to gas companies.

Following World War II, with the end of rationing of consumer appliances and industrial steel items, a boom in pipeline construction began that would last for the next twenty years. The rotary, high-pressure centrifugal pump had been developed during the war. When driven by high-pressure steam turbines, fueled by natural gas, it was well over twice as efficient as reciprocating pumps with more than 50 percent savings in operating costs. In Texas and Oklahoma, the Panhandle and Houghton fields were expanded. New discoveries were made in the Texas and Louisiana gulf coasts, which now accounts for about 40 percent of all U.S. natural gas discoveries. By the mid 1960s, the post-war boom in pipeline construction was complete. Natural gas was available in all U.S. states with the exception of Hawaii. In 1948 a milestone was reached; for the first time, the amount of natural gas consumed by residential customers exceeded the amount of natural gas vented or flared in production of petroleum.

The expansion would have been even faster and more complete except for the old problem of storage. With the new centrifugal pumps, the main transmission pipes could be enlarged to initially twenty-six-inch, then thirty-six-inch, and finally forty-two-inch diameter, where they seemed to have peaked. Under high-pressure, a large volume of gas could be stored in the pipelines, which would help solve the problem. The pressure could be gradually lowered during peak demands. Until large underground reservoirs could be found, however, the problem could not be totally solved and even larger pipelines would need to be built. These would run at 100 percent capacity in the winter and only 10 percent capacity in the summer, which would be prohibitively uneconomical. To level out summer and winter demands, the natural gas industry tried to market gas air conditioning and refrigeration, but electricity was winning the cooling market.

Deregulation

In 1954 the Supreme Court made a ruling, which became known as the Phillips Decision, which extended government control over wellhead pricing. The government now regulated all aspects of the natural gas industry, controlling the price on wellhead production and interstate pipelines. State government controlled the price on local distribution companies and end users. Even though the ruling was an attempt to extend the use of natural gas, the industry fought the Phillips Decision and for years tried to introduce legislation that would reverse it. They claimed the control of wellhead prices curtailed the exploration for new natural gas reserves.

By the late sixties and early seventies, the natural gas industry was in a decline. Electricity was winning the battle of consumer appliances. There was a perception of limited natural gas reserves, and even with a boom in housing during the late sixties, the gas industry was refusing to take on new customers. Events would soon change the image of natural gas as the "poor sibling" of the energy industry.

The first of these events was the Arab oil embargo in 1973. With the shortage of petroleum, industry and the government began looking around for other sources of energy that were not dependent on the mideast. The second was the extremely cold winter of 1976 and 1977, and the third was the environmental movement of the late seventies. During the winter of '76 and '77, many residential customers were restricted, or completely shut off, in their use of natural gas for heating. It now appears reserves were adequate for the cold winter, but some of the transmission companies had neglected to completely fill the reservoirs. This left some areas short during the heavy demands. Also, scientists were finding pollution problems caused by the massive use of petroleum and coal, and some were claiming natural gas was a clean alternative.

Congressional hearings were held in 1977 and 1978. These were the same hearings that James Schlesinger equated with time spent in hell and were forty years from the 1937-38 congressional hearings on the gas industry. The '37/'38 hearing had only two interest groups to contend with: the cities league and industry spokesman. The result was a simple policy that was meant to help both residential customers and the industry. In that forty years, however, the American public and American industry had learned the art of special interest lobbying. As a result, the '77/'78 hearings were prime examples of special-interest groups trying to go in all directions. It seemed each group had a computer and/or economic model to justify their direction. Most of the models seemed to forecast that despite an increase in natural gas price, there would be a dramatic increase in demand. This seemed to support the common belief in the industry that there is a "latent" demand for natural gas that was not being realized because of government regulations. The response by Congress was for deregulation. What followed was a somewhat confusing Natural Gas Policy Act of 1978.

The 1978 regulation essentially accomplished two things. The first was to create the Federal Energy Regulatory Commission (FERC) out of the old Federal Power Commission (FPC) and mandate it to reform natural gas pricing. Second was a consequence of the first, which was to deregulate wellhead prices, in effect to partially reverse the Phillips decision of 1954. It ended federal control of wellhead prices for newer wells but maintained price control over gas from vintage wells. The consequence of the legislation is perceived in different light depending on who is doing the observing. The natural gas industry states that production increased dramatically because of the perceived pent-up (latent) demand. A competitive market failed to develop, mainly due to the lack of incentives by the pipeline industry to purchase the most competitively priced gas supplies to sell to the local distribution network. The pipelines made the same profit, regardless if they purchase expensive or inexpensive gas. Producers produced more gas because they were now able to sell at higher rates but which the public was not buying because of greater cost. Industry claimed that pipeline regulation hindered the free market from developing. The "latent" demand, which all of the models predicted, did not materialize and led to a very large natural gas surplus in the early eighties. This would result in deregulation of the pipelines in later FERC orders.

It appears the gas surplus was simply because the tremendous increase in the cost of gas and the perceived shortage led to conservation measures by customers that dramatically reduced the amount of gas consumed. Because of the service cutbacks of the seventies and projected high costs, pipelines and end users contracted for long-term, large volumes of natural gas at very high prices. Some contracts even bound pipelines to pay for all the gas a producer could produce. In the early eighties, even though market prices for energy were declining, the price for older gas was held constant; therefore, the production from the older wells fell. The production of higher-priced gas from newer wells continued to increase under the long-term contracts. The industries' belief that the free market failed to develop because of the regulation of pipelines is probably not true. It is more likely that because the pipelines were tied to long-term contracts, they could not react to market conditions. This led to doubling of the revenues received by the producers from residential customers between the years 1978 and 1983. If conservation measures had not been instituted by the consumers, the revenue paid would have been approximately 20 percent higher.

FERC order 380, issued in 1984, outlawed provisions in a contract that forced consumers to pay for contracted gas even if they did not take delivery of that gas. This order allowed pipeline customers the freedom to find low-cost supplies and not be stuck with old, expensive contracts. In 1985 the FERC issued order 436 and established a voluntary program to encourage natural gas pipelines to allow open access to transportation of gas purchased by consumers from producers. This was meant to allow consumers to negotiate separately with producers and pipelines.

Later, in 1987, order 380 was modified by FERC order 500, leaving the buyer responsible for some part of the cost even if the product was not provided. The combination of orders 380, 500, and 436 was meant to balance supplies across production areas. If demand outstripped production in one area, producers from another area that had an excess of production could arrange transportation to where it was needed. The pipelines were owned by one group but could be accessed by other parties. The combination of the three acts also led to the establishment of gas marketing firms that had no ties to any one company and provided an intermediary service between buyers and other aspects of the industry.

The primary problem was that many gas producers were still supplying gas from the older regulated wells. This high-priced gas would in theory coexist with unregulated, more competitive gas. The sellers of the lower-priced gas would contract for transportation of the cheaper gas with the pipeline owners who were trying to compete with the high-priced gas. It did not work, and as a result, in 1989, the Natural Gas Wellhead Decontrol Act was implemented to remove all remaining wellhead price controls as of January 1, 1993.

FERC order 636 (1992) required the unbundling of gas sales, transportation, and storage. This was basically an accumulation of previous orders going toward a free-market program for the buying, transporting, storing, and selling of natural gas by consumers. They were now allowed to contract independently with all aspects of the industry. Later FERC orders in the nineties and 2000 led to further refinements in the streamlining of the different aspects of the natural gas industry.

Industry spokespeople will point to savings by customers and the reliability of natural gas service in the past fifteen years as an example of how well a deregulated free-market service can work. In reality, it is not that simple.

The industry states that deregulation and competition saved consumers tens of billions of dollars. Outside observers and some in the industry will show this observation does not hold up. Most of the savings estimates are based on the cost of gas under the old long-term, high-volume, and expensive contracts. Most of the savings come not from added competition but from FERC order 380, which dissolved many of the older contracts. If competition had actually been responsible for the savings, it would have been a result of reducing the margins paid to suppliers per unit of gas sold. The greater profits for the industry would have come from the larger amount of gas sold. Indeed, profits have gone up, but the margins have not gone down. This seems to eliminate competition as the reason for the savings.

As for added efficiency and reliability, that can also be called to question. During the last fifteen years, the United States has seen a series of mild winters, nothing really to test the reliability of the system. Some evidence seems to indicate the opposite. The summer of 2001 brought a shortage of natural gas that fired the turbines generating electricity for the state of California. Whether this was a true shortage or perceived shortage brought on by greed is still open to question. Some of the latest information seems to point to Enron as inventing the shortage to increase profits. The 2001 California shortage does, however, beg the question of the reliability of natural gas supplies.

Liquefied Natural Gas (LNG)

On a world scale, there appears to be an abundance of natural gas. Most of it again can be found in the Middle East and petroleum corridor through the Balkans. Without pipelines, however, it is extremely inefficient to transport natural gas over long distances. As a result of the petroleum shortages in the seventies, engineers and scientists began looking for a way to transport natural gas by using tankers. They found that by cooling and compressing the gas below its evaporation point (-260^0 F, -160^0 C), the gas could be liquefied and transported much more efficiently as a liquid. In the past thirty-five years, many of the world energy exporter nations have installed LNG port facilities to export their natural gas, and many of the world energy importer nations have installed LNG port facilities to import LNG. The United States has only built four LNG facilities for two reasons. First, many of the local residents feel the technology is too dangerous and have banded together to prevent the building of these facilities in their neighborhood. Second U.S. suppliers felt there is an abundant supply of natural gas on the North American continent that would supply us for years to come. As a result of these perceptions, there has been no feeling of urgency to build LNG facilities.

For many of the same reasons it is difficult to determine the amount of petroleum remaining (see Appendix A), it is also difficult to determine the amount of natural gas remaining. From the best estimates available, it looks as if the supply from the North American continent is static at best and probably beginning to diminish. Most of what remains can be found in Canada with some in Mexico. On a world scale, however, there appears to be at least twenty years' supply left before we begin to peak, even with increased usage as petroleum supplies dwindle. How natural gas will fit into the U.S. future energy scenario will depend largely on our willingness to install LNG facilities. Natural gas is by far the cleanest of the fossil fuels and may be the best transition fuel during the coming petroleum shortages. Also, as we begin the conversion to a hydrogen economy, with minor modification, the manufactured hydrogen could be integrated into the natural gas infrastructure.

Electricity
Foundation of Modern Society
The Beginning

Perhaps no area of modern civilization has been invested with so much human ingenuity as the electrical industry. The individual whose genius first initiated the technology that led to the development of the electrical industry was an Englishman by the name of Michael Faraday. In 1791 Faraday was born into a family of ten children with a father who was a blacksmith. He had very little formal education, and at the age of fourteen, was apprenticed to a French bookbinder, who allowed him to read the books he was working on. It was here that Faraday began his process of self-education that would ultimately lead to our modern electrical world. Faraday's self-education was very deficient in the area of mathematics, but he would become one of the world's greatest experimental physicists. Faraday's initial discoveries were in the area of chemistry, where he began research into the chemical analysis of fuels and later discovered benzene, carbon tetrachloride, and carbon hexachloride.

In 1800 the Italian physicist Alessandro Volta invented the electric battery. At the time, no one really understood electricity, but there was a sneaking suspicion that electricity and magnetism were somehow related. Finally, in 1820, the Danish physicist Hans Christian Oerstad had an epiphany while on his way to lecture. Oerstad, instead of his planned lecture, connected a wire to the two terminals of a battery and placed a magnetic compass near the wire. The north-seeking arrow on the magnetic compass was deviated toward the wire, demonstrating that electricity passing through the wire somehow created a magnetic force that attracted the compass. Faraday, as well as most of the scientists and public in Europe, became fascinated by electricity and magnetism.

In his research, Faraday was able to reverse the experiment of Oerstad. He passed a wire near a magnet and found that an electric current was created in the wire. He continued this line of research and developed two brilliant concepts, which led to much of modern science and industry. The first was the concept of the field. Faraday realized there was an area of influence around a magnet or an electric current that stored and expressed the energy of the magnet or the current. This area of influence would somehow attract or repel other magnets or electricity that came within that area of influence. He did not really understand what or why this influence was there, but he called that area of influence a field. He then went on to investigate these fields and showed that electric and magnetic fields are just different aspects of the same phenomena, whatever that was. As a result of these investigations, Faraday went on to imagine the entire universe as being composed of a series of fields. Because he neglected mathematics in his self-study, Faraday was never able to mathematically describe the field. This work was left to the brilliant Scottish mathematician/physicist James Clerk Maxwell. Maxwell's field equations, based on Faraday's experimental work, is one of the principle foundations of modern science and electrical engineering. The work also led directly to Einstein's work in relativity, light, and gravity.

The second concept developed by Faraday led to the invention of the dynamo (generator). Previous to Faraday's work, the only source of electric current was the chemical battery. This greatly inhibited the practical applications for electricity. Faraday discovered that if a wire was passed through a magnetic field, an electric current was induced into the wire. This electromagnetic induction is the fundamental principal of the dynamo, which Faraday quickly developed. He then developed the electric motor, which is based on the principle demonstrated by Oerstad of a current in a wire developing a magnetic field. Faraday was one of the first scientists to grasp the importance of educating the public on scientific advancements, thereby generating support for further research. He would go on speaking tours demonstrating his discoveries to the British public. When conducting his demonstration in front of the British Parliament, he was asked what electricity was. Faraday was said to have replied, "I have no idea what it is, but have no doubt that one day you will be able to tax it."

Although Faraday's research led to the development of the products that make up the electric industry, he received no income nor took out any patents on his discoveries. He published all of his findings and left it to others to develop and patent the products. A Frenchman, Hippolte Pixii, after reading Faraday's papers, constructed the first practical electrical generator. Pixii used rotational motion between the magnetic field and wire coil instead of the to-and-fro motion used by Faraday. Faraday was the first to demonstrate and build a transformer; however, two British inventors, Gaulard and Gibbs, invented and patented the first practical transformers, which ultimately led to the use of alternating current (AC) in virtually all commercial power generation. Faraday also designed the first electric motor, from which he received no income.

Most of the technology was in place for the beginning of the electric industry by the second half of the nineteenth century. The primary application for electricity at that time was in the use of electric motors and the electric arc lamp. It would take Edison's development of the electric light to introduce the massive growth of the industry we see today. Edison inaugurated this growth with his Edison Electric Illuminating Station on Pearl Street in New York. The station was steam driven, supplying two hundred and fifty HP to a one hundred ten volt direct current (CD) underground distribution system. The first waterwheel-driven (hydroelectric) generator was installed in 1882 in Appleton, Wisconsin. The early generating stations were all low-voltage DC and therefore had a very limited service area. DC electric generators produce electricity that continually passes through the wire in only one direction.

The problem of the limited service area for low-voltage DC was solved by the invention of the transformer. The transformer allowed the use of high-voltage alternating current (AC) distribution lines for long distances and lower-voltage AC for residential and commercial use. The transfer of electricity is extremely inefficient at low voltage, as much of the electrical energy is lost in heating the lines. Alternating Current is simply electricity that flows back and forth in both directions in the wire. In an AC generator design, electricity is first generated in one direction and then in the other direction. One complete cycle means the electricity has gone back and forth in the wire one time. Sixty-cycle AC is electricity generated to go back and forth in the wire sixty times a second.

With the transformer, AC-generated power can easily be stepped up to high voltage for transfer and down again for local use. This was not the case for DC. Initially, however, this was the only advantage for AC. There were no AC motors, and the electric light had not yet been developed. The Westinghouse Electric Company of Pittsburgh purchased the American rights to the transformer, and Edison and Westinghouse developed a serious competition over what type of generators and distribution would win the battle for American electrical industry. This battle of the "currents" would become an ugly public spectacle.

Many of the early entrepreneurs began their work in the electrical industry with Thomas Edison. Edison was very bright, imaginative, extremely hard working, and worked better in groups than individually. He therefore surrounded himself with individuals who were also bright, imaginative, usually underpaid, and quite often not given the work to suit their talents. As a result, many of them left his company and went off on their own to make a mark in the electrical industry. One of the earlier individuals who left was Frank Sprague, who was trained as a mathematician and engineer at the U.S.

Navel Academy. He began working for Edison in 1883 and provided the mathematical background Edison lacked. Sprague's calculations led to major savings in the amount of copper required for electrical wiring and a lucrative patent for Edison. Sprague received no financial benefit from his discoveries nor was he given more freedom in his research, which he had requested. As a result, he left and within a month had patented a constant speed electric motor that the Edison's company was forced to pay for the rights to manufacture. Sprague's company later went on to become a major developer in the electric railway industry.

Perhaps the greatest loss to Edison was a young electrical engineering wizard from Yugoslavia. Nikola Tesla began his career at Edison's European telephone company, where his genius was quickly noticed. He immigrated to the United States and was given a glowing recommendation from Edison's European manager. Despite the recommendation, Edison offered him a routine job at eighteen dollars a week. Tesla proved gifted at solving problems for which he received no extra pay. He also saw the tremendous benefits of alternating current electricity and constantly argued with Edison to develop AC electrical generators. Edison was having none of it and felt Tesla's ideas were terribly impractical. Tesla left, formed his own company, and took out patents on products developed from his ideas on AC electricity.

In a stroke of genius, George Westinghouse purchased the patent rights to Nikola Tesla's polyphase system of alternating-current generators, transformers, and induction motors. These inventions by Tesla are the foundation of all modern electrical systems. They were a quantum leap above anything then on the market. These developments gave Westinghouse a tremendous lead in the power generation market and ultimately led to our national AC system. This was not to happen overnight, however, and initially confusion reigned in the electric industry.

By the beginning of the twentieth century, electric generating companies abounded in the United States with no attempt at standardization. As an example, Philadelphia had around twenty different generating companies with distribution systems of one-hundred-volt DC, five-hundred-volt DC, two hundred twenty-volt three wire DC, single-phase, two-phase, and three-phase AC with frequencies ranging from sixty, sixty-six, one-hundred twenty-five, and one-hundred thirty-three cycles per sec. None of these systems could be interconnected, and sometimes users had two or three separate lines to handle the various distribution systems. Over the following decades, through consolidation and determining that economy of scale was a good investment in the power industry, Westinghouse's three-phase sixty-cycle AC generators won out.

Still into the early part of the twentieth century, electricity was more of a novelty than a need for most households. At the turn of the century, only one in fifteen houses had electricity. By 1910 Maytag and Hotpoint electric stoves and washing machines began appearing on the market, but it was not until the mid-twenties when electric appliances began gaining in popularity. Since the fixed cost of making electricity available to urban America was extremely high, it would have to be covered by big business. They would need to build the electric plants, build the distribution systems, wire the buildings, and hire a large staff to maintain and operate the facilities.

The Rise of the Industry

Samual Insull, who for twelve years was Edison's personal secretary and later took over the Chicago Edison company, would be the driving visionary behind the electrification of the United States. He determined that fixed costs were high but operating costs were quite low after the initial investments. He cut electric rates drastically and was therefore able to add large numbers of customers. He determined that an electric generator yielded a much higher efficiency factor as the percent of time it was operated increased. He developed the base load system design in which one generating plant would handle all the primary electric demand and as the system required more electricity, other generators would come on line to handle the increased load. This procedure yielded lower costs for the consumer and higher profits for the investor by lowering fuel costs. All of the developments led to large central generating plants through economies of scale, a process that would continue through the 1960s. (Economies of scale is the concept that it is much more efficient and economical to produce a product in large quantities. The cost per unit of the product is reduced as the amount of the product produced is increased. This concept has been one of the driving paradigms in the electrical industry since the twenties, and the size of generating plants has increased dramatically.)

Economy of scale, however, was not the only reason for the development of the large generating plants. Initially, much like the petroleum and gas industries, the owners of the early electric generating companies began fixing prices in areas where they felt competition would be ruinous. Later, with the Sherman Anti-Trust Law making this illegal, it was found that mergers or consolidations worked just as well. These mergers continued over the next forty years until by 1930, three holding companies controlled 40 percent of the nation's generating capacity. Samual Insull was at the forefront of these mergers and would become one of the most powerful figures in the electrical industry. As the government began cracking down on monopolies in the thirties, he would be convicted of embezzlement and mail fraud.

Another aspect of the electric industry, also promoted by Insull, began appearing in several states, namely state regulation of utilities in exchange for exclusive rights to operate in assigned areas. These "natural monopolies" have been the service providers in most states from the 1920s to the present.

They eliminate competition and make it almost impossible to have a business failure, in return for the electric rates to be regulated by state boards. Over the decades, four different types of electric utilities have emerged:

- Investor-Owned Electric Utilities: Financed through private funds and selling to both retail and wholesale customers.
- Municipally Owned Electric Utilities: Owned and operated by the municipality in which they operate.
- Federally Owned Electric Utilities: Usually generates power at a federally built and owned hydroelectric power plant.
- Member-Owned Rural Electric Power Cooperatives: Provides electricity to its rural members.

Over the next fifty years, electric prices fell as electric utilities were organized that would either generate their own electricity or buy from a producer, if economy of scale dictated. Many of the municipal utilities and rural electric coops initially generated their own power, but as larger and larger generating stations came on line, they found they could buy electricity cheaper than they could generate electricity. Old abandoned generating plants are a common sight in many of the municipalities across the country.

One of the first and largest of the producers was the U.S. Government, who began the production of electricity with the Tennessee Valley Authority Act of 1933. Under the law, the government accepted the responsibility to adopt navigation, control flooding, produce strategic materials for national defense, relieve unemployment, and improve living conditions in rural areas. The Tennessee Valley Authority (TVA) was created and mandated to produce, transmit, and sell electric power. This was later followed by the Rural Electrification Act of 1936, which created the Rural Electrification Administration. The REA provided loans and assistance to organizations providing electricity to rural areas and towns with populations under two thousand five hundred. These population densities were thought to be too small for private companies to serve economically. Over the next thirty years, more legislation would follow that allowed the federal government to develop hydroelectric projects, which would control flooding and produce electricity to sell in other areas of the country.

Just as the government reacted to the large monopolies controlling the gas industry, so too did it react to the monopolies controlling the electrical industry. In the thirties, the government stepped in with major legislation to curtail the power of these monopolies. The first of these was the Public Utilities Holding Company Act (PUHCA) of 1935. The PUHCA allowed the Securities and Exchange Commission to break up the large trusts that controlled the nation's generating capacity. At the same time, the Federal Power Act was enacted to regulate interstate transmission of electricity. By the early forties, it appeared that private utilities had taken a beating and public utilities were quickly replacing them. Roosevelt's Administration had pushed through several bills to cripple the monopolies, including Insull's Middle-West Utilities. They built public dams that sold the inexpensive federal power to public cooperatives and municipal utilities. Even with the legislation and hydroelectric projects, however, the private utilities still held on to 75 percent of the nation's generating capacity.

Also during this time, as the industry became more standardized and the size of generating plants increased, various companies began interconnecting their generating facilities into an interconnected grid system. This allowed for increased reliability if one facility went down or if peak loads would increase in one area more than another. The interconnected grid also allowed for lower reserve capacity by any one producer. This interconnection process continued until ultimately, the interconnected grid covered the whole of the North American continent. The grid, known as the North American Power Systems Interconnection, is divided into four distinct electrically separate areas. These areas are the Electric Reliability Council of Texas (ERCOT), the Western States Coordination Council (WSCC), Hydro Quebec, and the Eastern Interconnect, which is comprised of seven interconnected regions east of the Rockies with the exception of Texas. All of the generating plants within each of these four distinct areas generate electricity that is in phase. These four separate areas can exchange electricity through high-voltage DC lines but are not electrically in phase.

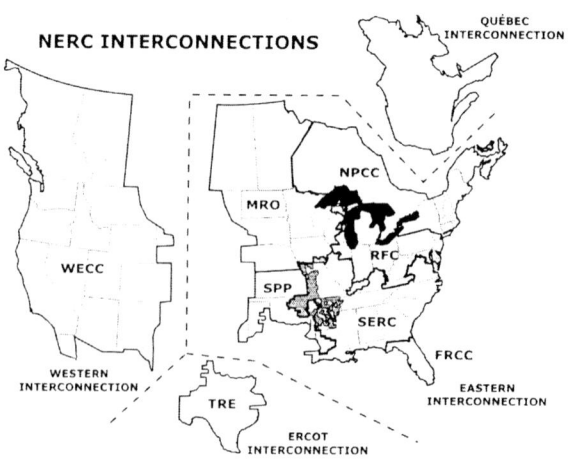

Fig. 4.1. A privately regulated organization, the North American Electric Reliability Council (NERC), is responsible for maintaining system standards and reliability. NERC coordinates the actions of the complete North American Grid System.

To be in phase, the electricity in the wires must be going back and forth at the same time. AC wires cannot be connected if the electricity in one of the wires is going in one direction while in the other wire it is going in the opposite direction. Since DC electricity is always going in one direction, it is much easier to interconnect DC electric lines of the same voltage than AC electric lines. Also, high-voltage DC will transport electricity much more efficiently than high-voltage AC; therefore, AC electricity is generated in virtually all generating stations, converted to very high-voltage with the use of transformers, changed to high voltage DC by the use of rectifiers, transported over long distances, converted back to AC through inverters, and finally reduced to a lower voltage by a transformer, where it can be used by the consumer. (Rectifiers and inverters were developed in the fifties as a result of the solid state (quantum mechanic) discoveries of the thirties.)

After World War II, all the legislation and technology was in place for the electrical industry to take off. The next twenty years would be considered the golden age of the industry, primarily for the investor-owned utilities. After taking a bit of a beating through the Roosevelt and Truman administrations, the IOUs made a remarkable comeback during this golden age of energy expansion in the United States. General Dwight D. Eisenhower's election in 1952 started a period of resurgent conservatism, which furthered the case for private industry over government-controlled electrical generation. The cold war was at its peak, and the public came to believe that private industry was the primary bulwark against the tide of communism. Republican members of Congress even tried to remove the inscription placed on federally constructed dams" "Built for the People of the United States."

Eisenhower considered the TVA an example of creeping socialism and gave several valuable leases, which the Truman Administration had reserved for public development, to investor-owned utilities. The investor-owned utilities played it to the hilt. Although they were protected monopolies (certainly not a good example of staunch capitalism since they completely avoided the free enterprise system), they used a brilliant public relations campaign to make it appear they stood alone against the communist menace. The gifted expert behind this advertising campaign was an individual by the name of Edwin Vennard. For fifteen years, from the mid-fifties to 1969, he would be the private utilities chief promoter.

He placed magazine ads, changed terminology based on public poll survey data, and manipulated public opinion to distrust anything that had to do with "big government." It was an extremely effective campaign. Support for public power decreased from 70 percent to 30 percent in only a few years. Vennard's campaign against municipalities, coops, and federal hydroelectric plants was perhaps one of the most misleading but effective advertising campaigns to that time. Many advertisers, political parties, and special interest groups learned from this campaign, and it is now quite common to see.

Because investor-owned utilities were much larger and could afford economies of scale they were able to block competitors. It was at this time that small companies and municipalities abandoned their smaller generating systems and began buying the less expensive electricity from the larger producers. Municipality generated electricity decreased from almost 50 percent to 10 percent over a forty-year period beginning in 1935. Private utilities and regional power pools, largely owned by private utilities, also restricted the supply of wholesale electricity through their control of the power lines. This process forced public power companies to buy power and eliminated any hope of decentralization of power production. This process, which began in the thirties, would increase in the fifties and continue into the seventies.

During the golden age of electricity, the investor-owned utilities were riding high. They had no competition, controlled most of the generating capacity in the U.S., had guaranteed job security, offered blue chip security stocks that paid high dividends, and could see no end to continued prosperity. In the early seventies, it did, however, come to an end. The bubble burst. Between 1973 and 1985, electric prices tripled. More than one-hundred eighty proposed power plants were canceled, and several in the process of being built were scraped. Strangest of all, in an industry where bankruptcy is unheard of, almost a dozen utilities were nearing bankruptcy. What happened, and can lessons be learned for the future from these events?

The Promise of the Atom, Etc.

After World War II, the United State entered into an area of supreme confidence. The Allies had won the war. U.S. energy supplies and technology had played a major role. We were the only country that held the secrets to the atomic bomb. In 1946 Congress instituted the Atomic Energy Act, which created the Atomic Energy Commission (AEC) to "conserve and restrict the use of atomic energy for the national defense, to prohibit its private exploitation, and to preserve the secret and confidential character of information concerning the use and application of atomic energy." Then on September 22, 1949, Truman announced the Soviet Union had exploded an atomic bomb. The announcement shattered U.S. confidence and completely eliminated the U.S. monopoly on nuclear technology. Truman ordered the construction of the hydrogen bomb, and the arms race was on. In 1951 Congress amended the Atomic Energy Act and began sharing atomic secrets with allies in NATO. Truman, however, still felt nuclear secrets were "too important a development to make the subject of profit seeking." Then in 1952, the Eisenhower Administration was elected and the promotion of nuclear energy as a fuel to generate electricity was placed high on the agenda.

Originally the AEC was not confident in the use of nuclear fuel in reactors. Most of the top scientists who worked on the Manhattan Project left after the war, and the nuclear reactors they built to enrich the nuclear material used in the bomb were falling apart. No less an individual than Robert Oppenheimer felt there were too many economic and technical problems in the development of nuclear reactors for producing electricity. In a 1947 report, he stated, "It does not appear hopeful to use natural uranium directly as an adequate source of fuel for atomic power."

With the election of Eisenhower, however, the commercial application of nuclear energy became a top priority. His state department argued that commercial atomic power could be used in global foreign policy gains in the cold war against the Soviet Union. On December 8, 1953, Eisenhower gave his Atoms for Peace speech in front of the United Nations, in which he stated that atomic energy should be taken out of the hands of the military and used for peaceful purposes by bringing energy to the world. Although leading Democrats of the time felt the AEC should construct and monitor nuclear reactors, under Eisenhower and a Republican Congress, private corporations won out and were encouraged to build nuclear reactors. (In hindsight, the Atoms for Peace program was one of the worst foreign policy acts ever developed by the United States. Not only did it require the American taxpayer to develop nuclear industries for other countries but at present time, when nuclear proliferation is the single most serious challenge to world peace, the U.S. has only been able to retrieve about one-sixth of the nuclear waste produced by the Atoms for Peace program.)

Most of the technology used in the development of nuclear electrical generators came from research the U.S. Navy was developing for use in nuclear submarines. In 1947 Admiral Rickover assembled a team of gifted engineers to build the nuclear navy. Rickover was a political and organizational genius and after seven years of planning and research, in January 1955, he delivered the first nuclear-powered submarine. In its first trials, the *Nautilus* broke all previous Navy records by traveling ten times further than any previous submarine; a total distance of one thousand three hundred miles underwater.

Because of the success of the nuclear submarine technology and the slow pace of developing commercial nuclear reactors, the AEC officials borrowed the naval technology to build their first demonstration project. Rickover's demonstrated "light water reactor" was used for the AEC supported joint project between Westinghouse and the Duquesne Light Company of Pittsburgh in 1954. With an opening speech by President Eisenhower, construction of the first nuclear-powered generating plant was started at Shippingport Construction Site along the Ohio River.

(Three years after that event, I remember my junior high science teacher telling us students that electricity would soon be "too cheap to meter," a sentiment expounded by the government and the media at that time.)

Although the government was strongly promoting the private use of atomic energy, many, if not most, utilities were quite wary of the promotional hype. Nuclear generators were much more expensive than other fuel-powered plants, there was no guarantee of long-term access to atomic fuel, the reliability factor was unknown, and safety risks could be extremely high. Very few businessmen were willing to invest in such a speculative venture. To overcome this hesitation, in 1955, the AEC commissioned a Power Reactor Demonstration Program. The program supplied costs and engineering data on the construction of plants, a guaranteed supply of fuel, and research grants to utilities and manufacturers.

Still the utilities were not buying. They were extremely worried that with a plant failure, a major accident could occur that exceeded available insurance coverage. The AEC then commissioned a study to determine the probability of such an occurrence and the damages that could result. When the study was completed, it stated that a worst-case accident is highly unlikely but if it did occur, three thousand four hundred people would die immediately and another forty-three thousand would be injured. Property damage would cover a forty-three thousand-square mile area and result in seven billion dollars' worth of liabilities—hardly a report to instill confidence in investors.

Nuclear promoters determined that more extreme measures were needed and went back to the federal government. They were able to pass a measure that absolved utilities of most liabilities, even if they were responsible for the accident, and provided five hundred sixty million dollars to cover any damages—certainly not an investment in the free enterprise system. The government also reduced any regulatory constraints that might slow the development of the technology and create problems with licensing the plants. The AEC became more of a promoter for nuclear energy than the monitoring agency that it was originally intended. If any questions concerning the reliability of new plants were raised, they were largely ignored in favor of quickly completing construction.

Also, to help promote nuclear energy, the Eisenhower Administration subsidized the international market for nuclear equipment. American taxpayers supplied four hundred seventy-five million dollars for the Euratom project. The friendly European countries involved were to use American equipment and nuclear fuel to build one-thousand megawatts of nuclear-generating capacity in Europe. In 1962, the AEC stated they spent over one and a third billion dollars in taxpayer money on research and development into nuclear power plants; more than twice the investment of private industry.

While post-World War II America was a great time for the expansion of private utilities, it did not turn out as well for manufacturers of electrical equipment. General Electric (formally Edison Electric Light Company and Thomson-Houston Electric Company) and Westinghouse had become the giants of the industry. For a period of thirteen years, however, from 1946 to

1959, they and many of the smaller companies (a total of forty-seven) had been fixing bids when selling equipment for most projects in the country. The TVA (Tennessee Valley Authority) had been aware of the bid fixing for several years by noticing that virtually all of the bids they received during that time were nearly identical. Because TVA had been branded an example of "creeping socialism" by the Eisenhower Administration, they laid low and did not raise any flags. By 1959 they were fed up and awarded a hydroelectric generator contract to a British firm. All hell broke loose. Westinghouse, General Electric, and Allis-Chalmers (third largest manufacturer) accused TVA of being unpatriotic and driving American workers to the food lines. They stated the only reason the British firm could underbid them was because of the lower wages they paid their workers. TVA showed documented evidence this was not true and countered by stating the American bids were high because the American companies were fixing their bids, thereby eliminating competition.

As a result of these charges, the federal government began investigating the bid fixing. The results showed a pattern of collusion and corruption that extended throughout the industry. Prices of equipment and tools were controlled and raised well beyond a fair price for manufacturing and marketing. If smaller companies refused to go along and offered a fair bid, they were frozen out of any future bids. By the end of the investigation in 1962, over two-thousand damage suits were entered against the electric equipment manufacturers and many of the country's top executives were sent to prison. As a result of this fiasco and under a new Democratic administration, the government was losing some of its love affair with private industry. U.S. taxpayers had been supporting companies that had been cheating U.S. taxpayers for years.

By the early sixties, Westinghouse and General Electric felt they had received as much from the government as they could. The utilities were still not buying, even with all of the previous government promotion. Westinghouse and GE needed the nuclear power market to boost sales and change their image after the bid-fixing fiasco. Their new chief executives finally decided to make the utilities an offer they could not refuse, a completely finished nuclear power plant. In 1963 General Electric offered utilities a completed power plant at a set price in which General Electric would assume all the building responsibilities. These responsibilities included managing all the subcontractors, labor problems, delays, and cost overruns, with the utility simply turning the key to start the plant.

The Oyster Bay plant was the first "turn-key" nuclear power plant built for the Jersey Central Power & Light Company. It was a technical and public relations success. The reactor, at five-hundred fifteen megawatts, was three times larger than any previous nuclear power plant and performed at expectations. To build it, however, cost twice as much as expected, but

General Electric felt they could recoup their losses in future plants as the technology became more standardized. Utility interest picked up and Westinghouse and General Electric offered twelve more turn-key plants. The savings through standardization never materialized and the companies lost an average of seventy-five million dollars on each contract, for a total of almost one billion dollars—a huge amount in 1966 dollars, but the rush was on. Reactor orders went from seven in 1965, to twenty in 1966, to thirty in 1967. With the turn-key offer, the utilities quickly changed their attitude toward nuclear electrical generation from one of skepticism to acceptance in a very short time. This change of attitude in spite of no long-term operational data for nuclear generators was the beginning of their problems.

A Rude Awakening

In the mid-sixties, small cracks began to appear in the electric utilities' world view. Just how reliable was this huge infrastructure, this grid for transmitting the nations power? Were the new power plants more efficient than the older ones? Was economy of scale the correct paradigm for the industry? Would the demand continue to increase? Would electricity really become "too cheap to measure"? These and other questions would challenge the industry over the next thirty years, questions that continue to the present, with answers that will determine our nation's electrical energy future.

The first of the cracks that appeared was the blackout in Northeast United States on November 9, 1965. It covered eight states and Ontario, a total of eighty-thousand square miles, and lasted almost thirteen hours. Most of the affected population persevered with stoic humor, but the affected utilities were not laughing. A chink had appeared in their armor. Consolidated Edison investigated the accident for over ten years. Con Ed's chairman then made a statement guaranteeing such a blackout would never occur again. Three days later, on July 13, 1977, another blackout occurred that affected nine million people for up to twenty-five hours. This time, the affected population did not react with such stoic humor, and chaos reigned over much of New York City, with rioters and looters causing fifty million dollars in damage. These and other power outages forced utilities to invest large sums into upgrading existing facilities and distribution systems.

Since 1977, there have sporadic blackouts throughout the United States and Canada. In the summer of 2001, California suffered from a series of blackouts and rolling brownouts. Parts of the state did not feel any effects from the shortages, while other areas had daily problems due to energy shortages. It now appears the brownouts/blackouts were the result of a terrible marketing and purchasing agreement agreed to by the state and not a result of technical problems.

In August of 2003, the largest blackout ever in the United States again occurred in the Northeast. It appears the blackout started in Ohio and cascaded through much of the eastern U.S. and Canada. Again, it looks as though the transmission system was outdated and there should be large investments into the national grid system. How this investment will be paid for will be the topic of debate over the next few years.

The second crack in the utilities' armor appeared in 1962, with Rachel Carson's publication of *Silent Spring*. The book began questioning the effects of major industrial growth on the environment. Eight years later, the first Earth Day was held and many in the United States were beginning to feel that bigger may not be better. Polls showed that pollution was now the second most important concern of the public. Citizens reacted in various ways. Some challenged growth through legal battles; others cut down high-voltage towers; others instituted legislation (Edmund Muskie was able to push the National Air Qualities Act through Congress); some used civil disobedience; and many cut back on their energy use. Utility officials initially tried to address the concerns by simply planting more greenery around their plants. It didn't fly. They were finally forced to install pollution-control devices on all existing power plants and address environmental concerns on proposed power plants.

The third crack and perhaps the most important for future concerns was the whole concept of economy of scale. The whole idea of the larger the generating plant the more efficient the operation was no longer true. Around 1967, the peak size for maximum efficiency had been reached. From that point on, any new plants that came on line would increase the cost of electricity. That increase added to the cost of upgrading the infrastructure, and introducing pollution-control measure certainly put an end to the idea of "too cheap to measure"; in fact, the average cost of electricity increased almost every year from 1967 to the mid-eighties and more than doubled over that time period.

The fourth but not the final crack in the utilities' armor was the Arab oil embargo of 1973. Although utilities only used 13 percent of the nation's total consumption of oil, the embargo forced the increase in price of all fuels, including coal. This increase in fuel costs added to the price of generating electricity. With the increases in electric rates, many consumers instituted conservation measures that reduced their electric consumption. Many utility executives did not perceive the reduction in electric consumption as an omen but only as a mild aberration. With the oil embargo on, the Nixon Administration was pushing ahead with "Project Independence," which included the construction of two-hundred nuclear power plants. The utilities were listening.

The final massive crack in the utility paradigm was that of the nuclear reactor. After the intense marketing campaign, first by the government and then by the equipment manufacturer, most of the utilities were jumping headfirst onto the nuclear bandwagon. Much of this jumping, however, was based on inadequate information. It now appears the government and the manufacturers had not been completely upfront with the costs of building and maintaining a nuclear power plant. In 1965 power companies predicted that one-thousand nuclear-powered generating plants would be on line by the year 2000. In reality, no new orders were placed after 1977, and in 1985, only eighty-two nuclear plants were in operation.

The utilities will give all kinds of excuses for the failure of the nuclear industry, primarily blaming environmental concerns and governments regulations. Under closer scrutiny, these excuses do not hold up. The regulations were not imposed until the earlier reactors started breaking down due to improper construction or materials. The cost of a new reactor rose tenfold between 1965 and 1985. Cost overruns were endemic, sometimes by a factor of fifteen times the original estimated cost. It is a very complex technology, with over forty thousand valves going into a large reactor, ten times the amount going into a comparable-sized coal-or-oil-fired generating plant. Complex technologies are expensive and prone to failure.

With all of the economic problems the nuclear power plants were facing, nuclear energy was most likely doomed; however, the final death knell to the industry were the accidents at the Three-Mile Island and Chernobyl generating plants. There had already been several earlier accidents, both in the U.S. and abroad, in nuclear reactors; however, these incidents were not made public. The Three-Mile Island accident, which occurred on March 28, 1979, was the most publicized, the largest to that time, and had the potential for being cataclysmic. The Nuclear Regulatory Commission's final investigation stated that the reactor had been within one hour of a catastrophic meltdown and the release of major amounts of radioactive gases into the air was avoided by "dumb luck." Never again would nuclear energy be looked on as the savior of our energy needs.

The Chernobyl accident would not have the "dumb luck" that Three-Mile Island escaped with. Following a safety check in which the plant was operated above design parameters, at 1:23 A.M. on April 26, 1986, two explosions occurred at the unit four reactor of the Chernobyl Nuclear Power Plant in the Ukraine. The reactor and connected building were destroyed in the explosion, while nearby buildings were ignited by burning graphite. Radioactive emissions spread across western Soviet Union, eastern Europe, and eventually the whole of the Northern Hemisphere. The fires burned for several days until the core melted. About 70 percent of the melted fuel (one-hundred thirty-five metric tons) remained uncovered for ten days until cooling took place.

One hundred thirty-five thousand people were immediately evacuated from within a thirty-km radius exclusion zone. A total of four hundred fifteen settlements were later evacuated and the residents resettled. The radioactivity released was estimated to be more than two hundred times that released from the combined nuclear explosions at Hiroshima and Nagasaki. Over eight hundred thousand people were involved in the cleanup, and millions more were exposed to the radiation. The death toll from this exposure will never be completely known. Green Peace estimates about thirty-two thousand individuals were killed as a result of the explosion. Other estimates are lower with a few being higher. In nearby areas, thyroid cancer has increased two hundred fold, but the final estimate of the damage will not be known for decades with an increase in other cancers and birth defects beginning to appear.

In an extremely pathetic attempt to confound the public, the Uranium Information Center, which seems to be the information arm of the World Nuclear Association, issued in 2007 a Nuclear Issues Briefing Paper #22, which claims as of May 2004 there were a total of only fifty-six casualties as a result of the Chernobyl meltdown. At the same time, a much more impartial estimate came from an information, communication, and networking platform, provided by Switzerland and the United Nations. The report stated that a total of one hundred thousand cases of thyroid cancer in people of all age groups would result from the Chernobyl accident.

Where Do We Go from Here?

The utilities industry has traditionally been one of the most subsidized industries in the history of the United States. Almost every aspect of their existence is subsidized, from the fuel they burn, to their low-interest loans, to paying few or no taxes. Public-owned utilities receive tax-exempt municipal financing and low-interest federal loans, and most pay no property tax and no federal income tax. Some are required to offer free public services, up to 10 percent of their income, in exchange for not paying taxes. They also have preferential access to low-cost federal hydroelectric generated electricity.

The private-owned utilities certainly are no shining example of the free enterprise system. They have lobbied for incentives that have allowed almost a fourth of private utilities to pay no taxes. Like other businesses, private utilities are able to write off one-half of their long-term financing costs and can depreciate any new investments in fifteen years. Investments in nuclear power can be depreciated in ten years. Specifically for the utility industry, the IRS allows a charge to customers for taxes they may never have to pay. In 1982 the one hundred fifty largest utilities received five billion dollars in tax payback. The total of the tax breaks to utilities cost each household in the U.S. more than four hundred dollars per year. Utilities argue that government interference costs them money in excess regulations and environmental controls. As we have seen, this charge does not hold up under closer scrutiny. To be allowed to operate as "natural monopolies," however, the utilities have agreed to government regulation.

In the late seventies, all this began to change. To avoid the need for new power plants because of the increase in peak demands, to encourage industrial users to generate their own power during these peak demands, and to encourage the generation of electrical power by the renewable technologies, the federal government passed the Public Utility Regulatory Policies Act of 1978. The act forced the electric utilities to buy power generated by industrial customers and renewable energy sources. The result was to open the door for a whole new class of electrical generation, co-generation, or non-utility generation, introducing competition for the electric utilities for the first time in their existence.

The price at which the utilities bought the power was and is the main bone of contention, both by the utilities and the power producers. Southern California Edison was forced to pay fifteen cents/kilowatt hour for electricity generated from solar powered sources, when it was only three cents/kilowatt hour to buy it wholesale from traditional sources. The solar technology people argued there were environmental costs in the electricity from the traditional sources that were not being considered, and it would also take time to reach an "economy of scale" in the renewable technologies. In other areas of the country, the utilities were only required to pay "avoided costs" for the electricity they were required to purchase from renewable sources. The utilities would set this avoided cost structure based on their own data. They would then buy electricity from a wind system producer for 1.2 cents/kilowatt hour and sell it down the road for 7 cents/kilowatt hour. This made it impossible for many of the early small wind system companies to survive.

For the industrial users generating their own power, there was decrease in natural gas costs and development of smaller more efficient turbine generators. This allowed manufacturers to install generating plants that could produce electricity cheaper than they could purchase from their utilities. During the 2001 blackout/brownouts in California, many of the manufactures could earn more profit from selling their power than from producing their products. Several quit manufacturing and sold all of the generated electricity. Utilities claim these new producers do not have to include infrastructure investment in their pricing so they could undersell the utilities.

These avoided costs (called stranded costs) are the primary discussion topics in the electrical industry to this day. The federal government passed the Energy Policy Act of 1992, which made it much easier for the wholesale generator to produce non-utility electricity. The natural gas transmission system was opened to greater access at this time—so too was the electrical transmission system opened to wholesale suppliers.

Today there are more than three thousand electric utilities of which two thousand are public utilities that generate 14 percent of the country's electricity. There are nine hundred co-ops that generate 8 percent of the country's power while the two hundred investor-owned utilities generate 76 percent of the electricity. Six federal utilities generate the remaining 2 percent. Coal-fired power plants produce about 50 percent of the power while natural gas generates around 10 percent. Nuclear power still provides 14 percent but will decrease as the older plants are retired and no new plants are coming on line. Petroleum fuels about 10 percent of the power plants. Hydroelectric dams supply about 3 percent of the nation's total demand with other solar technologies supplying 10 percent. The solar technologies are the fastest growing of all the sources and as the cost of the technologies decrease, they should grow even faster.

What will happen to the utilities and the various types of generating plants under the new guidelines and environmental constraints is still an open debate. They were established under conditions that made perfect sense at the time. Today there is a two hundred billion-dollar market at stake and fierce competition for the customer. Can the utilities survive in this competitive market? What will be done about the stranded costs, and who will maintain the infrastructure of the system if the utilities fail? With the new technologies, will we even need the national grid system to the extent it now exists, or will it gradually die out? We will revisit this question again in chapter seven, when we look at possible future energy scenarios for the electric industry.

CHAPTER FIVE

It's All About Energy

In discussing the energy situation with others or giving talks to general audiences, the authors find it surprising that most of our public are simply unaware or refuse to accept that we are at that juncture in our history at which we have to change our way of doing business. Most seem to feel more petroleum will be found or there will be some magic discovery that will solve our problems and we can carry on as usual. There is simply no evidence to support this. In fact, all the evidence is to the contrary, and we must change to a more efficient way of life. In this chapter, we will look at the various aspects of the energy process and examine what is available to us in terms of energy sources and technologies for transitioning to a sustainable future.

To better understand the energy industry and further simplify the world energy problem, we can divide the science and technology of energy into five processes for finding, converting, and using energy and look at each of them separately. These five processes include:

1. Finding and extracting some form of crude energy that is available. For our purpose, we will define crude energy as any energy that is not in a usable form, e.g., crude petroleum, tar sands, oil shale, solar rays, and wind currents.

2. Converting that available crude energy into a form that can be stored, transported, and/or consumed.

3. Storing the energy until it is needed.

4. Transporting the stored energy to where it is needed.

5. Converting that stored energy into applied mechanical work or heat.

Sometimes these processes overlap. For example, as the solar energy is extracted, it is usually converted to usable forms, such as heat or electricity, all in one process. As we discussed in the previous chapters, a tremendous amount of human ingenuity has gone into developing these energy processes, primarily in the fossil fuel and electrical industries.

When these different processes are discussed, two scientific concepts, that of *efficiency* and *energy density* will be used quite often. Further discussion on thermodynamics and the mathematical derivation of the concept of efficiency in thermodynamics are given in Appendix B. Very simply, however, efficiency is the ratio of how much energy or work you get out of a process over how much energy or work you put into the process. Multiply this ratio by one hundred, and we get a percentage of energy out over energy in. For example, an efficiency of 10 percent means that for every ten units of energy we put into a process, only one unit of useful energy is extracted from the process. The other nine units of energy either cannot be used or are expelled in the form of waste heat. Sometimes, in a process defined as co-generation, that expelled heat energy can be used for a constructive purpose. (In some of the modern literature, the word, "regeneration" is used instead of "co-generation.") Examples of co-generation are heating your car in the winter with the waste heat from your engine or using waste heat from an electrical generating plant to heat nearby buildings, preheating water, various industrial processes, and improving the efficiencies of generating electricity.

To determine the overall efficiency from extracting the crude form to applying the energy for a practical use, the efficiency of each individual process is multiplied by the efficiencies of the rest of the processes. For example, the efficiency of a wind turbine for converting kinetic energy from the wind to mechanical energy may be 50 percent. The efficiency of converting the mechanical energy from the turbine to electrical energy for the electric generator on the wind turbine may be 90 percent. The transmission of the electricity to where it is needed is around 95 percent, and the conversion from electric to mechanical energy in powering an electric motor may be 90 percent. Multiplying all of these efficiencies together gives an overall efficiency of about 38 percent. If we were to store the electrical energy and retrieve it before use, we would then need to include the storage and retrieval efficiencies to achieve overall efficiency.

Energy density is the amount of energy you can store in a given volume of space or given mass of material. For instance, natural gas stored at normal temperature and pressure has a relatively low energy density. When the gas is pressurized, the energy density is increased and if the gas is liquefied, there is a dramatic increase in energy density.

In this chapter, we will examine all of the current sources and technologies available for each of these processes, the efficiencies of each process, and the energy density of the stored fuel and try to determine if this points to a direction for a future world energy economy. The first of these is finding and extracting a form of crude energy that is available to us. We can subdivide all of the currently known crude energy forms into three basic sources: forms of solar, non-solar, and fossil (fossil was, at one time, biomass solar).

Step 1: Extracting the Crude Energy
Types of Solar Energy

Solar energy in its various forms comes either directly or indirectly from the sun. A list of these includes:

Direct Solar: Using direct solar radiant energy.

Wind: Solar energy creates the world wind patterns through uneven heating of the earth's surface.

Hydroelectric: Solar energy creates the energy available in the falling water through evaporation and condensation (rain) of the world's oceans and other bodies of water.

Biomass: Solar energy grows the plants used in biomass fuels (wood, alcohol, biodiesel, and crop residue). Also, in some local areas, animal waste is used as a fuel.

The technologies for extracting crude solar energies are well developed. For direct solar, we use passive or active solar panels, concentrating collectors, and photovoltaic cells. A passive panel is usually a large window facing south (in the Northern Hemisphere), that allows the sun's radiation into the building to help heat it in the colder months. The active panel usually has a collector plate, which gets hot. Some type of fluid (either gas or liquid) is passed over the heated plate by the use of a pump or fan and transfers the heat energy to where it is used. Both passive and active panels convert the solar radiation directly to heat, which is used on site, either in space heating or water heating. The concentrating collectors will focus the solar radiation to produce higher temperatures and are used either for direct heat or for producing electricity through photovoltaic cells or turbines. Photovoltaic cells convert solar radiation directly to electricity. All of these technologies have been proven, and the efficiencies are increasing while the cost is decreasing. The efficiency of direct solar is usually in the 10 to 20 percent range. The primary reason for the low efficiencies is the inability to collect the total solar radiation over the wide range of frequencies that strike the earth.

Another form of direct solar energy scientists have been researching over the past thirty years is the extraction of energy from the temperature gradient in the ocean or in special energy ponds developed for this purpose. As the sun heats water, the upper layer of water will become warmer than the water at lower levels. If the water in the special ponds is covered with a dark surface, the temperature between the lower and upper levels will be amplified even more. This temperature difference (gradient) can be used to power a heat engine using a working fluid that evaporates and condenses at a lower temperature than water. As the temperature gradient becomes sharper, the heat engine becomes more efficient.

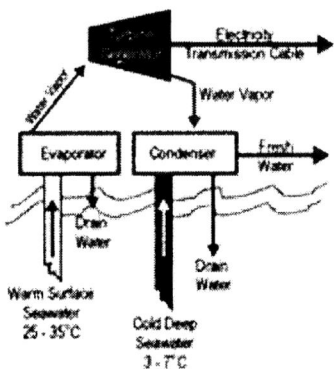

Fig. 5.1. Diagram of thermal gradient heat engine. Warm surface water is taken into the evaporator, where the working fluid is evaporated and expands, passing through the turbine and back to the condenser. There it is condensed by cooler water and returned to the evaporator, and the process is repeated.

Wind has been used for centuries for pumping water and grinding grain. In the last several decades, the technologies have been developed and improved, so it is a very reliable, low-cost source of crude energy. Through a wind turbine-driven generator, the kinetic energy of the wind is converted into electricity. The theoretical maximum efficiency of a wind turbine is 59 percent. If all of the kinetic energy is extracted from the wind, the wind behind the turbine would be traveling at such a slow speed that it would interfere with the incoming wind. Theorists have determined that at least 41 percent of the energy must remain in the processed wind in order for it to get out of the way of the incoming wind. The efficiencies of most wind systems will vary depending on wind speed, usually in a range from 20 to 40 percent. Wind is now the fastest-growing solar technology and one of the fastest-growing energy sources in the world.

Hydropower has also been used for centuries throughout the world, primarily for grinding grains. In slightly more than a century, most of the useful sources for large-scale hydropower projects have been harnessed for hydroelectric generators. Electricity is generated by falling water passing through a turbine, which turns a generator or alternator. The amount of electricity generated is a function of the height of the falling water and the amount of water falling. The efficiencies of water-driven turbines have increased over the past decades and now approach 80 percent. Many of the older hydroelectric sites, especially those in high-erosion areas, are filling up with silt and are unable to generate the amount of electricity for which they were designed. This may decrease the amount of hydroelectric energy available to us in the coming decades.

Biomass, primarily in the form of wood, is another crude source of energy that has been used since humans first discovered fire. Over the past century, with the rise in world population, the world's forests have been decimated. As a result, except in localized areas where forests have been protected or areas maintained as woodlots, wood is no longer an option as a fuel source.

In the last few decades, some scientists and agricultural communities have been pushing alcohol produced from fermentation of crops as a viable alternative fuel for the internal combustion engine. Some scientists claim it takes more energy to grow the crop and produce the alcohol than the energy we extract as a motor fuel. In other words, the process has an overall negative efficiency. Supporters of the process state that when using the byproducts, mash (a byproduct of the distillation process) as stock feed and fiber as bedding, the process becomes more efficient and develops a positive efficiency. Improvements in the fermentation process over the past few years, along with the latest research, seem to indicate there is a slight energy gain, about 10 percent in the overall process. The final judgment, however, is still out on the use of agricultural-produced alcohol as a fuel source. The process is certainly not able to replace gasoline on a world scale without seriously impacting food supplies, which even now are in danger of becoming inadequate to supply all of the world's needs.

Alcohol production from grasses and trees is now being researched. This looks much more promising in that it does not affect the food supply. The overall efficiency of transferring from raw material to alcohol is much better (no intensive farming for production). At this time, the technology is not on line, and the process for breaking down the large grass and wood molecules must be found before they can be fermented to alcohol. The technology looks about ten years down the line but looks much more promising than conversion of crops to ethanol. The energy density of grasses is low, but the process could be used in some local areas.

Another bioproduct being pursued across the world is biodiesel. Many of the beans and seeds grown around the world that are now converted to cooking oils can also be used in diesel engines. Much of the research in Asia and the United States is in the use of palm oil and soybeans as a biodiesel fuel. In the long-term, this would affect world food supplies but in some areas, it may be used as a transition fuel.

Crop residue and animal waste are two sources of crude energy that could be used on a local scale. Most crop residue is primarily carbon, which can make a good fuel by burning for direct conversion into heat. In many cases, if the crop residues are left in the soil, the nitrogen in the soil will be consumed during the decomposition process. The nitrogen must be replaced by fertilization so there may be some benefits to removing the residue before it decomposes in the field and use that residue as a fuel. Again, one has to look at the energy balance and the long-term effects on the soil to determine if this process is viable.

Animal waste can be used either through direct combustion of the dried waste or through anaerobic decomposition. Anaerobic decomposition is used in many urban waste treatment plants and some livestock yards as a method to produce methane (primary ingredient of natural gas). An extremely rich nitrogen fertilizer is the byproduct of the decomposition process. Many dairy and beef farmers in Europe and some in the U.S.A. are producing methane in anerobic digestors, storing it in large expandable structures, and producing electricity by combusting the methane in turbines. They then sell the electricity to power companies during peak demands for premium rates. Again, these processes could be used on a local scale.

Non-Solar Forms of Energy

The two main forms of non-solar energies are the nuclear technologies and tidal energy. The nuclear energy sources are the result of one of two processes: fission or fusion. Fission is the splitting apart of a large nucleus with the loss of mass in the process, and fusion is the combining of smaller nuclei into larger nuclei with the loss of mass. The lost mass is converted to energy, described by Einstein's famous equation: $E=MC^2$.

Since Three-Mile Island and Chernobyl, the nuclear fission industry in the United States is about dead. With the new safety designs, insurance, cost overruns, and nuclear waste disposal problems, nuclear fission plants are prohibitively expensive. There have been no orders for the past several decades. Many scientists felt the breeder reactor would solve most of these problems and again make fission plants a viable alternative. In the 1990s, government funding for research into breeder reactor technology was eliminated which, for now at least, doomed the fission industry. Even if more fission plants were to come on line, it would be only a short-term solution to the world energy needs. The crude fuel for the fission reactors would run out within a century.

Fusion is the reaction that powers the sun and the thermonuclear (hydrogen) bomb. At one time, fusion was the darling of the nuclear energy enthusiasts. There is very little radioactive waste generated in the fusion process, and there is a very long-term supply of crude energy for a fusion reactor. The problem is in controlling the reaction. There has been research into the peaceful use of fusion for almost half a century, and we still appear to be a long way from practical applications. If there is ever a breakthrough in fusion technology, it could play a large part in future energy economies and could perhaps be one of the magic solutions that many are seeking.

Tidal energy is produced through gravitational interaction of the sun, moon, and Earth's oceans. Two methods are used to harness tidal energy. The first is a turbine, much like a wind turbine, is placed below low water level in an active tidal area. As the tide comes in, it turns the turbine, extracting energy. The turbine is turned 180^0, or a second turbine is installed opposite the first to extract the energy as the tide goes out. In the second method, a suitable location must be found, which includes a bay that can be isolated by an artificial barrier (wall) from the ocean, sea, or whatever body of water the bay is on. When the tide is coming in, the barrier is closed, allowing the water level to increase to a predetermined height along the barrier. When the height is reached, a water door in the barrier is opened, allowing water to funnel into the bay, turning a turbine connected to a generator much like a hydroelectric dam. When the tide goes out, the barrier door is again closed, the turbine is rotated 180^0, and the process is reversed. The bay will remain filled while the tide goes out. When the tide is low enough, the door will open and the water will again pass through the turbine. The method is ingenious. Lately, however, there have been concerns by environmentalists over the barriers changing the ecosystems of the bays through which they are constructed. Also, the technique is extremely site specific and has limited application.

Fossil Fuels
Oil Shale and Tar Sands

Various forms of fossil fuels were discussed in chapters two, three, and four when looking at the growth of our present energy economy. These forms include coal, petroleum, natural gas, propane, and butane. The primary reason the world economy has developed around the fossil fuels is that in a primitive technological society, nature has accomplished several of the processes needed to make to make fossil fuels a desirable form of energy. Nature has transformed dead organic material into a usable form of energy, which can be stored and transferred to where it is needed quite easily. To extract a desirable form of energy from the stored fossil fuel, all one has to do is burn it. As the Industrial Revolution came of age, more elaborate methods of extracting energy from the fuels were developed.

The two areas of fossil fuels that have not been discussed and that many claim will be the saviors of the petroleum industry are oil shale and tar sands. We will briefly examine each of those here to see if those claims are valid. Pitch tar, asphalt, and heavy oil are all names for similar products. Natural petroleum products occur in a continuous range of thickness, from normal light crude which flows quite easily, to much denser, more viscous crude, which will not flow at all. Normally, the lighter the crude, the more expensive and desirable it is. Tar sand deposits are of the extremely thick variety and probably did not form at an adequate depth to produce the lighter crude. This heavy crude is primarily made up of multiple ring carbon molecules and to upgrade them into the desirable lighter molecules, hydrogen atoms must be added to the carbon. Also, at normal temperature, the tar cannot be removed from the sand. The primary reason for the tremendous interest is that there is estimated more oil in Canadian and Venezuelan tar sands than in all the world's conventional oil wells combined.

In Canada three-fourths of the heavy crude now used is obtained by surface mining with extremely large equipment. This method is very investment intensive and traditionally dependent on natural gas to remove the heavy oil from the sand and enrich the oil with hydrogen atoms so it can be piped to refineries. The other method for removing tar from the sand is to heat the heavy oil below ground with steam. Once it gets warm enough to begin flowing, the oil is removed through either the same pipe used to distribute the steam or with a second pipe inserted for that purpose. In this process, the initial investment is much lower but even more dependent on natural gas to generate the steam and enrich the heavy crude with hydrogen molecules.

The tar sands in Venezuela face most of the same drawbacks as those in Canada with the added problem of containing large amounts of vanadium. Vanadium is a heavy metal that forms deposits on turbine blades and promotes the formation of sulfuric acid and the corrosion of the blades. At this time, Venezuelan heavy oil is being distributed by adding water along with detergent to help promote an oil/water emulsion. If the mixture ratio is at least 70 percent heavy oil, 30 percent water, the emulsion can be transferred through pipelines and be burned directly as a fuel.

The problem in extracting heavy crude is that all processes are extremely heavy consumers of energy. With the diminishing supplies of natural gas, if heavy crude is to play a part of the world's energy future, other methods must be found for extraction and enrichment of the petroleum. Some have proposed using solar or nuclear energy as a heat source to extract the heavy crude and water as a source of hydrogen to enrich heavy molecules. A strong argument could be made to just use the solar or nuclear as a direct energy source or to generate hydrogen and forget about the tar sands with their associated problems. Much more research and thought is needed to determine which path is best to fit the world's energy needs.

At this time, there is some petroleum being extracted from Canadian tar sands using surface-mining techniques. About 10 percent of Canada's shipment to the U.S. comes from the Tar Sands in northern Alberta. Much of the local native population is in an uproar because of the environmental impact on the land. Besides the adverse effect on the land, the extraction and conversion also produce large amounts of nauseous vapor and carbon dioxide. It is too early to determine if this process will continue but right now, with the high price of petroleum, it looks very attractive to many investors, irregardless of the environmental consequences.

Many of the same problems associated with the extraction of heavy oil from tar sands can be said about the extraction of heavy crude from oil shale. It is still very attractive because of the large amount of energy available, and the U.S. (in the western states) has about 60 percent of all the world's oil shale. Oil shale is a newly formed rock that contains petroleum and again was formed too shallow to produce the lighter form of crude. All one has to do is heat the rock to drive off the petroleum. The heavy crude from oil shale has very high sulfur content and contains paraffin, which quickly turns the extracted crude back into a slimy solid as it cools. Like all heavy crude, the molecule from oil shale is carbon rich and must be enriched with hydrogen molecules. This again is accomplished with natural gas or water, both of which are in short supply. Also, oil shale creates a disposal problem with the leftover shale. When it is heated, the shale expands 20 percent over its original volume, making disposal difficult. Again, oil shale may play a major role in our energy future if the problems can be worked out and depending on what future energy path we choose. As we will see in the next chapter, however, it would be better for our world and biosphere if we began the transition from fossil fuels and did not bother with the tar sands and oil shale.

Step 2: Converting Crude Energy into a Usable Form

The next three steps in the energy industrial process are the crux of our energy dilemma. In the last one hundred years, humans have found it is very easy to find crude forms of energy. It is sometimes difficult to convert these crude forms into a more usable form and usually quite difficult to transfer and store these usable forms to a time and place where we can apply them to our local needs, e.g., heating, air conditioning, transportation, and agricultural and industrial processes.

We should stop here and look at an area of science that has developed along with the Industrial Revolution. That is the science of thermodynamics, which is the study of energy. Two fundamental laws govern all of thermodynamics and basically all of nature: the first and second laws of thermodynamics. The first law simply says that in any energy conversion process, you can never get more energy out than what you put in. In other words, you cannot develop a process for converting from one form of energy to another, that is more than 100 percent efficient. There is no free lunch. The second law states that in any process for converting from one form of energy to another you can never get as much energy out as what you put in. In the process of conversion, some energy will be lost, and the process can never be 100 percent efficient. Even though the science of thermodynamics is well developed and no violation of the first and second laws has ever been found, there are always those who claim to have the perfect machine. It seems as though every few years, some inventor will claim to have developed a perpetual motion machine or a machine that produces more work than the amount of energy put into it. None of these claims have ever been verified.

The primary reason for the evolution of petroleitus has been the ease that crude petroleum can be processed to a more usable form, with a high energy density, and the ease with which that usable form can be stored and transported. In chapter two, we saw the development of the petroleum industry along those processes. Let us now look at other possibilities for storing and transporting other usable forms of energy.

In the previous section, we looked at the converting of solar energy to usable forms. These included direct solar, wind, hydroelectric, and biomass. Although wind was originally used to pump water, the primary use by modern technology for both wind and falling water is the generation of electricity through turbines. Hydroelectric is a fairly continuous supply of electricity and can be transferred over long a distance through power lines. Wind and direct solar are intermittent forms of energy, and we must be either able to store them or have some backup source when the wind is not blowing or the sun is not shining. Direct solar can be used either to generate heat or be converted to electricity through photovoltaic or concentrating collectors.

There are two methods that architects, engineers, and scientists have developed for direct conversion of solar energy into heat and storing the solar energy as heat. The first of these is called thermal mass storage, and the second is phase change storage. Both are used for local application, e.g., houses or commercial buildings, and the heat energy stored is not easily transferable over even short distances. Thermal mass storage is the heating of a large mass with some heat source, quite often a solar system, either passive or active, and drawing the heat from storage when the sun is no longer available. Large masses can store a very large amount of heat when just being heated a few degrees. These have proved quite successful, and other intermittent sources of energy are now being used to heat them. Thermal masses are usually some form of masonry/stone construction or a large volume of water. Refractory bricks are now being used and heated to very high temperatures, usually around 1600^0 f.

When building our house approximately fifteen years ago, the authors installed a swimming pool that is primarily used as a thermal mass. The pool contains twenty-two thousand gallons of water, and when heated to eighty degrees, will heat the house for almost two weeks with a drop of only ten degrees in pool temperature. The authors use waste wood and passive solar for heating the pool. Most power companies in the northern United States are now offering special rates for heating a thermal mass during off-peak hours and using the stored heat during peak hours. This will be discussed in greater detail in section three.

Fig. 5.2. Winter picture of authors' house. Super insulated with indoor pool for thermal storage.

Fig. 5.3. Swimming pool in authors' residence. Twenty-two thousand gallons of water will store enough energy to heat the house for several weeks. The pool also has recreational value.

One drawback to this method of storing heat energy is the large area (volume) needed to install a thermal mass. The phase change method is an attempt to solve this problem. Any substance will store a large amount of heat when changing phase from a liquid to a gas or a solid to a liquid. The substance will give up that heat when the process is reversed. The most common phase change substance is a eutectic salt called sodium sulphate decahydrate. This is a fairly common salt that has the property of melting around 90° F, which makes it a nice medium for storing heat energy. When energy is available, the salt is heated and the molecules become exited and break the molecular bonds. The energy is stored in the difference in molecular movement between a solid and a liquid. It has a much higher energy density than the thermal mass technology. When the source of energy is removed, the molecules will slow down and the salt will return to a solid form, releasing the energy. The phase change heat storage method is not as widely used as the thermal mass method but could gain favor as the technology becomes more available. To get optimum use from either of these methods, it is extremely important for the buildings in which they are used to be well insulated.

Thermal mass and phase conversion technology can also be used to cool a building. With the use of a heat pump/air conditioner system, the storage system can be cooled during off-peak hours and during peak hours, air can be passed over the cool mass and distributed throughout the building.

Electricity is generated through one of two processes: either a chemical/electrical process (batteries) or a mechanical/electrical process (generators and alternators). As mentioned in chapter four, the process for the mechanical/electrical generating of electricity was developed by Michael Faraday in the mid-nineteenth century. Very simply, electricity is generated in a wire that is passed through a magnetic field or a magnetic field is passed over that wire. To generate the electricity, all that is needed is the relative motion between the wire and magnetic field. In the first case, you have what is called a generator and in the second case, you have what is called an alternator. In a generator, the magnetic field is stationary and mounted on the outside casing of the generator. The inner rotating component that passes through the field is composed of the wires in which the electricity is generated. In an alternator, the wires are mounted on the outside casing and the magnetic fields are mounted on the rotating component, which then rotates past the wires generating electricity. A force (mechanical energy) is needed to pass the wire through the magnetic field or vice versa. That mechanical energy can be supplied by a heat engine (internal or external combustion) or turbine powered by some form of stored or crude energy, e.g., fossil fuel, wind, or water. The efficiency of conversion from the crude energy to mechanical energy can be as low as 10 percent with heat engines to 80 percent with modern turbine technology. Modern generators or alternators can be quite efficient at converting mechanical energy to electrical energy, usually in the range of over 90 percent. The overall efficiencies when going from crude forms of energy to electrical will range from less than 1 percent to as high as 70 percent.

Step 3: Storing the Energy

We have already discussed the various methods of storing fossil fuels in the forms of kerosene, gasoline, natural gas, etc., and also of storing solar energy as heat. Because most of the crude forms of energy can be directly converted to electricity, the storing of electricity is perhaps the single biggest challenge to modern science and technology. All of the crude solar energy forms can be converted to electricity through photovoltaic, heat engines, or some form of turbine-driven generating system. The problem becomes what to do with the excess electricity during times of surplus generation and the question of where we get the energy in times of low production. In short, the transfer to a sustainable economy is dependent on our ability to store energy. It is the holy grail of modern energy technology.

Unless the solar/wind electrical system is specifically designed for remote application, virtually all of the present wind turbines and photovoltaic systems are installed to work in conjunction with the electrical grid. When the wind is blowing or the sun is shining, electricity is being generated and pumped into the grid. The company or individual who installed the generator will receive the financial benefits for the electrical energy generated and someone on the grid will use it, not necessarily the group that installed the system. When the wind is not blowing or the sun is not shining, the grid will compensate for that loss. The grid becomes a storage mechanism for the solar/wind system. This is a very handy method of distributing the energy and allows for interested groups to install the solar/wind electrical generator in a location that is most advantageous for the wind or solar system. This has worked well up to the present time but as more solar/wind systems are installed, most likely (and hopefully), the situation will occur that in a time of peak production there will be an excess of energy being pumped into the grid. If we can develop adequate electrical storage, this excess can then be used to supply electricity during time of low production.

An electrical generating plant is more efficient and economically feasible if it can supply a continuous level of electric energy. Peak demands on that energy usually occur in the morning and evening. Very low demands occur during the night. The most advantageous situation is if the generators in a generating plant can somehow generate a continuous supply of electricity and during times of low demand (off peak) store the electricity and then use that stored energy during peak demands. Over the past century, some electrical utilities have developed ingenious methods for storing electrical energy. One method is pumped hydroelectric storage. During low-demand hours, when there is an excess of generating capacity, water is pumped from a lower reservoir to an elevated reservoir using the excess electricity. During peak demand hours, water is released from the reservoir, passing through a water turbine and generating electricity. The reservoirs can be either natural or artificial and are becoming more common as the demand for electrical storage is increased. The storage efficiency for this method is around 64 percent. Of the energy used to pump the water to the higher level, approximately 64 percent is recoverable.

Compressed air storage is being investigated by several U.S. power companies, and one has been installed in Germany. In this process, during times of excess production, pressurized air is pumped into large underground caverns with the use of a compressor and electric motor/generator. During times of peak demand, the pressurized air is released from the cavern through a turbine, which rotates the generator/motor producing electricity. The efficiency of this process is less than 40 percent. The low value is primarily because of the heat generated by the compressor when the air is pumped into the storage. As any gas is compressed, the molecules that make up the gas become more and more constrained in the distance they can move in any one direction. This increases the speed of the molecules, which is an expression of heat energy. Some of the heat energy stays in the compressed air, but much of it goes into heating up the compressor. As the temperature of the compressor increases, some process must be used to cool the compressor. Engineers and scientists are searching for methods to store that heat energy gained during the cooling process and use it for some constructive purpose.

Also, as was mentioned in the previous section, thermal mass and phase change technology is being promoted by the utilities for use in residential and commercial buildings as a form of heat storage during off-peak hours.

Another method of storing electrical energy that has been investigated over the past several decades is that of flywheel storage. Any child who has played with a top has seen a good demonstration of flywheel storage. To store the electrical energy, an electric motor is used to spin some mass. The amount of energy stored is proportional to the mass being rotated and the square of the speed at which the mass is rotated. It is therefore more important to achieve a very high rotational speed than it is to have a large mass. The limit to the maximum rotational speed is the speed that the mass will fly apart. Flywheel storage technology has been used to power busses and automobiles. Heavier, stronger composite materials are now being investigated and hold some promise for the future.

At this time, the most widely used method to store electrical energy is in a device commonly known as a secondary, rechargeable, or storage battery. There are various types of rechargeable batteries, but all of them work on the same principle. The battery is composed of two terminals inserted into an electrolyte. When electricity flows out of the battery, a chemical change takes place in the terminals and the electrolyte either uses up the active materials or deposits inactive material over the active material. When electricity flows into the battery, the process is reversed and the battery again becomes charged. Over a period of time, the electrolyte will be consumed or the deposited material on the terminals will harden and battery is considered "dead." The most common type of rechargeable battery is the automotive or recreational lead acid battery. It is very reliable, low cost, and can be recharged over a large number of cycles. The primary problem is that it is very heavy and has a very low energy to weight ratio. The nickel hydride is also popular but more expensive and only has a slightly higher energy to weight ratio. Some of the newer batteries are much lighter and have four to five times the energy to weight ratio of the lead acid battery but are very expensive, unreliable, and can only be recharged a few times. Research into new types of battery storage is continuous and could be very profitable if a relatively reliable, inexpensive, lightweight, and high energy to weight ratio battery were developed. Most hybrid automobiles are now using a lithium ion battery, which seems to hold great promise for the future.

Lately the most researched and perhaps favored type of electrical storage is that of hydrogen storage. Hydrogen, as a gas, is the most abundant element in the universe. On Earth, however, free hydrogen is almost nonexistent. Except for a very small percent, about one part in a million in the atmosphere, hydrogen exists only when combined with other elements, primarily oxygen in water (H_2O). A large amount of free hydrogen is produced worldwide for use in various industrial and agricultural processes. The current primary method of production is through a catalytic reaction between fossil fuel, primarily methane (CH_4), and steam, which produces hydrogen and carbon dioxide. The hydrogen produced in this manner is not pure and has small amounts of carbon mixed with the hydrogen. As we will see later, this can be a problem for certain applications of hydrogen.

Water is by far the largest source of hydrogen on the earth, and electricity can be used to separate the oxygen from the hydrogen. The hydrogen can be stored for later use and quite easily transferred to other locations. The improvement in technology of converting electricity to hydrogen has increased the efficiency of the conversion process over the past several decades from around 70 percent to around 90 percent. Hydrogen can be stored as a gas at standard pressure but has an energy density only about one-third the energy density of natural gas.

There are several methods for storing hydrogen. One is in standard-bulk storage tanks as in propane storage. The energy density at these pressures is extremely low, which makes this type of storage almost cost prohibitive and not terribly practical. Underground cavern storage is being investigated;

however, the small hydrogen atom may allow for excessive leakage. A promising method for hydrogen storage is in natural gas pipelines. This will be discussed in the next section.

The favored methods of hydrogen storage being investigated are metal hydride storage, high-pressure storage, and liquid storage. Metal hydrides are metals with the ability to absorb hydrogen like a sponge when the hydrogen is placed under a small pressure in the presence of the hydride. When the hydride is lightly heated, the hydrogen is released. It is an extremely safe and fairly inexpensive method for storing hydrogen. The primary problem with metal hydride storage is the heavy weight and fairly low energy density. An area of research that may improve the energy density problem is in the use of carbon or graphite nano tubes or composite fibers as a storage sponge.

High-pressure hydrogen storage is the favored method being researched by the automotive industry. With present technology, five thousand pounds per square inch (psi) is the maximum pressure available for storage. The auto industry considers the energy density at this pressure is still too low for practical use in automobiles. They are shooting for a pressure tank that will store hydrogen at ten thousand psi and feel this will offer a practical automobile that will be able to travel a distance comparable to that traveled by modern vehicles. It takes quite bit of energy to store gases at these pressures, which lowers the efficiency of the storage method. The efficiency of the system would be improved if the energy used to pressurize the gas can be retrieved as the gas is withdrawn from the tank. With the newer types of vehicles researched, this retrieval of the storage energy is very possible.

Liquid hydrogen storage has been used for several decades in the space program. Hydrogen will liquefy at -253° C and has an energy density of one thousand times the energy density as a gas. As a liquid, hydrogen has a much higher energy density than any liquid fossil fuel, and the technology for storing it has been well developed by NASA. Drawbacks to liquid storage are the expense, possible hazards, and quite a lot of energy expended in the liquefying process. During the energy crunch of the seventies, the commercial aeronautics industries researched the use of liquid hydrogen as a jet fuel. As more fossil fuel was made available, the research was dropped.

With the interest in hydrogen as fuel, a new type of battery called the fuel cell is being researched as a source of electricity. In a normal battery, there is a very finite time of use until the battery is fully discharged. In a fuel cell, as long as there is a continuous input of fuel, usually hydrogen, there will be a continuous output of electricity. The conversion efficiencies of most fuel cells is around 60 percent, which means if it is used in conjunction with an electric motor to drive vehicles, the overall efficiency will be around 54 percent. The space program has been using fuel cells for the past forty years but in the past decade, several other companies have committed their resources to the research.

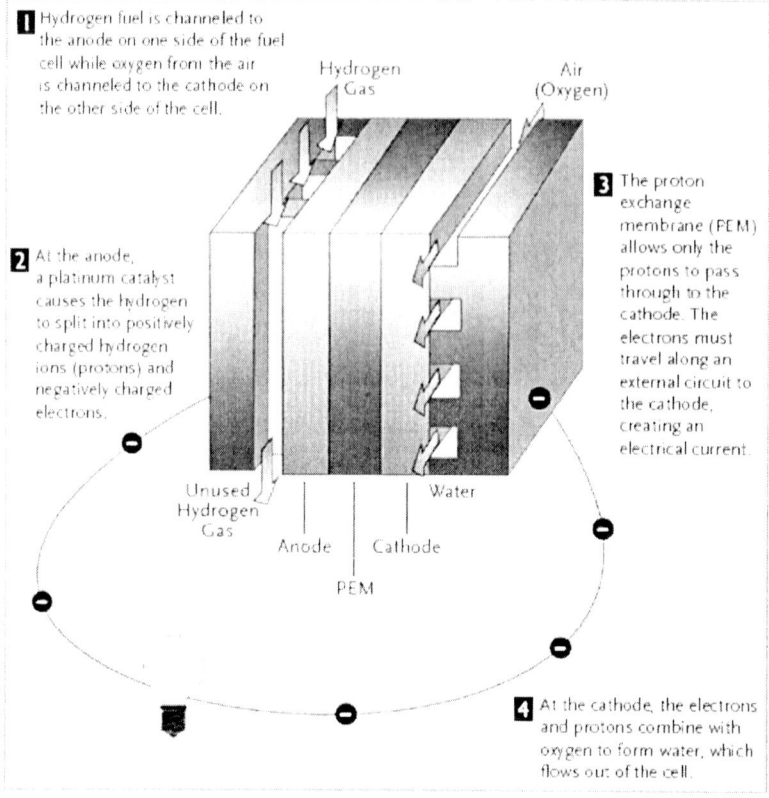

Fig. 5.4. Diagram of fuel cell. Several different types are being researched around the world.

The major auto manufacturers are investing large sums of money into fuel cell technology. Many scientists and engineers are quite certain that with the decline in fossil fuel sources, the hydrogen production, storage, transportation, and fuel cell technology will be the next major technology base for the world economy. At this point, several design and demonstration projects have been developed. Honda now has a working model of a hydrogen-powered, fuel cell car that reportedly compares very well with the performance of most internal combustion engines.

There are several problems yet to be solved for large-scale application of this technology. The first is the cost of the fuel cell. Virtually all fuel cells need platinum as a catalyst, which makes them quite expensive. Much of the research is in decreasing or eliminating the amount of platinum needed in fuel cell construction. A second problem is the fouling of fuel cell by impurities, primarily carbon contained in the fuel. This limits the life of the cell. The hydrogen used at this time is produced by stripping hydrogen from fossil fuels. This process leaves small amounts of carbon molecules mixed in with the hydrogen. The problem could be solved by producing pure hydrogen through electrolysis from water. The third problem is simply there is no infrastructure for supplying the hydrogen at stations similar to the present gas stations. The technology is available and certainly is a solvable problem. It will, however, take a commitment from government and industry that is not yet in place.

Below is a table showing the relative energy densities of various fuels and batteries, both in energy per unit volume and energy per unit mass.

Material	By Volume	By Mass
Diesel Fuel	10,700 Wh/l	12,700 Wh/kg
Heating Oil	10,400 Wh/l	12,800 Wh/kg
Gasoline	9,700 Wh/l	12,200 Wh/kg
Butane	7,800 Wh/l	13,600 Wh/kg
LNG (-160°C)	7,216 Wh/l	12,100 Wh/kg
Propane	6,600 Wh/l	13,900 Wh/kg
Ethanol	6,100 Wh/l	7,850 Wh/kg
Methanol	4,600 Wh/l	6,400 Wh/kg
250 Bar NG	3,100 Wh/l	12,100 Wh/kg
Liquid H2	2,600 Wh/l	39,000 Wh/kg
150 Bar H2	405 Wh/l	39,000 Wh/kg
NiMH Battery	280 Wh/l	100 Wh/kg
Li-Ion Battery	200 Wh/l	150 Wh/kg
Lead-Acid Battery	40 Wh/l	25 Wh/kg
STP Propane	26 Wh/l	13,900 Wh/kg
STP NG	11 Wh/l	12,100 Wh/kg
STP H2	3 Wh/l	39,000 Wh/kg

STP: Standard Temperature and Pressure
Wh: watt hours
Sources:
http://xtronics.com/reference/energy_density.htm
http://hypertextbook.com/
http://www.ior.com.au/ecflist.html
http://www.batteryuniversity.com

Step 4: Transporting of Energy

Usable energy is transported in the form of gas, liquid, solid, or as electrical energy. Solid and liquids are usually transferred by tankers, trucks, or by rail. Gas (because of its low-energy density) is usually transferred through pipelines. Electricity is usually transferred over high-voltage AC or DC lines.

In some cases, liquids will also be transported through pipelines when extremely large amounts of energy are to be transferred from one point to another. In chapters two and three, we saw the rise of the petroleum, natural gas, and electrical industries because of the development of the methods to economically store and transport energy. Any new energy source should have the ability to be stored and transported over long distances in an economically feasible form.

Heat storage in the form of thermal mass and phases change are usually designed into a building and not transferred over any distance. The one minor exception to that is when phase change storage is used in a heat pump or air conditioning system. A heat pump is a device for transferring energy from one point to another, usually over very short distances. In a heat pump, a working fluid in the gas form (usually one that evaporates and condenses quite easily) is pumped under pressure to a reservoir from which it can remove heat energy, turning the working fluid from a liquid to a gas. The energy is removed by slowing down the molecules of the reservoir, which means cooling the reservoir. The reservoir can be at a much lower temperature than the final temperature desired in the heated area. The working fluid is then circulated back to the pump, where it is compressed and liquefied, removing the heat energy through the phase change from a gas to a liquid. If the process is reversed, the heat pump becomes an air conditioner.

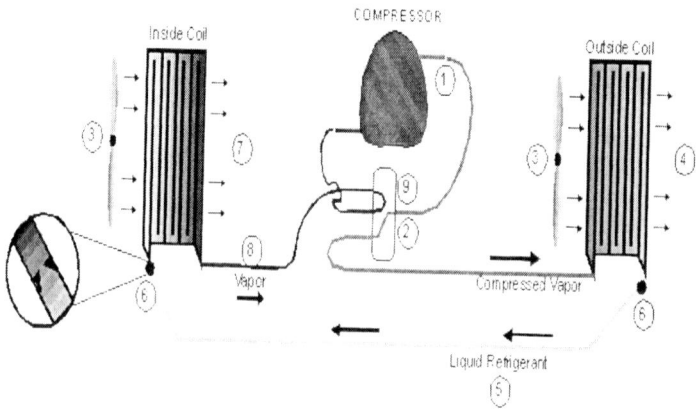

Fig. 5.5. Heat pump/air conditioner. The reversing valve 2/9 reverses the flow of the working fluid. In cooling mode (diagram), line 8 is the vapor going into the compressor and out line 8 as a compressed vapor and converted to a liquid and giving off heat energy to the outside in coil 4. The refrigerant passes as a liquid to inside coil 3, where it is vaporized, absorbing heat from the coil. In heating mode, the reversing valve is switched thereby, reversing the process.

The heat pump does not produce any new energy. It simply removes the heat energy from a reservoir by lowering the kinetic energy of the molecules and transfers that energy to where it is desired. It consumes energy, usually electrical, to run the pump that compresses and moves the working fluid from the heat reservoir to heated area. The ratio of the heat energy moved over the electrical energy consumed is called the coefficient of performance (COP) of the heat pump. The COP of a heat pump can be as high as five. Five times as much heat energy can be transferred from a reservoir to a desired location as energy is needed to produce that transfer. As the temperature of the reservoir decreases, the COP of the heat pump falls. At about 20° F, the COP falls to around one, which means the heat pump will do no better than electrical heating elements.

Hydrogen looks very promising as a method for transporting energy and can be transferred from one location to another through natural gas pipe lines. Although hydrogen has only one-third the energy density of natural gas, because it is less viscous than natural gas, it can pass through the pipelines much easier—in fact, three times easier. This allows for the same energy flow to pass through the pipelines with hydrogen as with natural gas.

Hydrogen can also be transported under high-pressure or as a liquid when used as a fuel for vehicles. As a pressurized gas, the energy density will be quite low for use in an internal combustion engine that only gets 10 to 20 percent efficiency. The new fuel cell technology, however, makes hydrogen look very promising as a vehicle fuel.

Step 5: Converting Stored Energy into Applied Energy

Basically all methods of converting stored energy into some form of applied energy can be divided into three processes. The first is through some form of chemical reaction, the second is through the expansion of gas, and the third is converting the energy into electrical energy. The electrical energy is then converted into heat by heating elements or mechanical energy by an electric motor. The process of converting all fossil fuels and most biomass to an applied form of energy is through the chemical reaction of combustion, which then produces an expansion of a gas. Nature has stored this energy in the fossil fuels in the molecular bonds. One can think of the energy in these bonds as similar to the energy in a compressed or stretched spring. During combustion this energy is released in the form of heat and expansion. After combustion this energy can be used directly or converted to another form, usually mechanical energy, through some form of heat engine and then into electrical energy if so desired. During the overall process of going from molecular bonds to heat to mechanical to electrical, some of the energy will be lost during each stage and dissipated as other forms of energy, e.g., heat or sound. The efficiency of the overall process can be anywhere from 10 to 35 percent depending on the type of engine used.

Heat engines can be of three types: internal combustion, external combustion, or turbine. The internal combustion heat engine is used in virtually all of the world's ground-based transportation systems and initially used to drive most of the electrical generators. Put very simply, through some valve/timing mechanism, fuel is inserted into a firing chamber of an internal combustion heat engine and the fuel is ignited. Once the fuel is ignited, the combusted gases will expand, forcing a piston downward, producing mechanical energy. The expanded gases are exhausted, new fuel is inserted, and the cycle is repeated. In both cases, internal and external combustion, the motion of the piston is converted into rotational motion by the crankshaft.

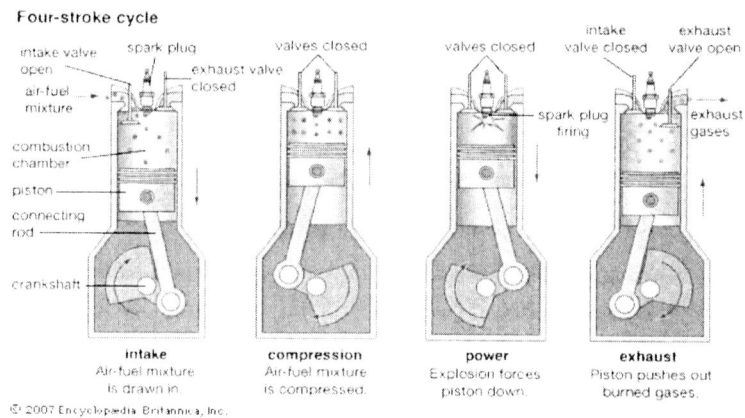

Fig. 5.6. Diagram of working four-stroke internal combustion engine.

In an external combustion heat engine, some type of working fluid is heated externally and expanded into the expansion chamber, which forces the piston to move. The expanded fluid is then cooled back to a liquid, and the cycle is repeated. The Industrial Revolution was initially built on the steam engine, which is a very inefficient (around 10 percent) external combustion engine. The primary advantage of the external combustion engine is that they have the ability to use just about any fuel available to heat the working fluid. An old/new type of external combustion heat engine, called the Stirling external combustion engine, looks very promising for future use. The modern Stirling engine uses nitrogen as a working fluid, has very few moving parts, and can use just about any heat source as a fuel. Because it has very few moving parts, it is very reliable. Some Stirling engines have been operating for twenty years with no maintenance. Because of design consideration, small Stirling engines less than 10 HP are much easier and cheaper to manufacture. They should be very popular in any new energy economy.

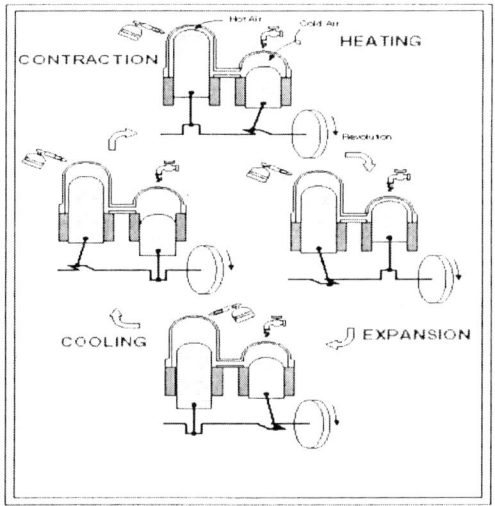

Fig. 5.7. Diagram of working stirling engine.

The most efficient way to drive a generator or alternator is through a turbine. It is a set of blades that extracts energy from a moving fluid, either a gas or a liquid. The turbine blades are designed so that a moving fluid will strike the blades in such a manner as to produce rotation of the turbine. In a turbine heat engine, a continuous supply of fuel in injected into a combustion chamber along with compressed air. The fuel and air mixture is continuously ignited and expanded through a turbine before it is exhausted through some exhaust mechanism. The rotating turbine blades convert the heat energy into usable mechanical energy that also drives the air compressor for the combustion air.

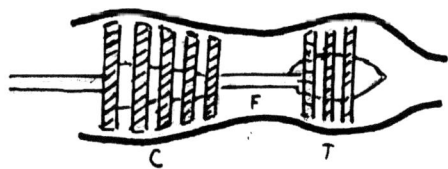

Fig. 5.8. Air is drawn into the turbine through the front opening (left) by the fans (c). Fuel is continually injected into area f, where it is combusted, creating a high-pressure area, which drives the exhaust gas past the turbing blades (t), causing them to rotate.

The *determining factor* for the efficiency of a heat engine is the allowable temperature difference between the fuel or working fluid entering the combustion/expansion chamber and the exhaust gases expelled from the chamber. With modern materials the maximum efficiency obtainable for a heat engine is around 30 percent. For most internal combustion engines used in the transportation industry, however, the expected efficiency is around 10 to 15 percent. The external combustion engine is usually less, around 5 to 10 percent. Over the past half-century, the turbine heat engine has replaced the external combustion heat engine with dramatic increases in efficiencies, from around 10 percent to around 60 percent. The turbine achieves these efficiencies by passing a cooling fluid through the turbine components allowing for higher combustion temperatures or using heat from the exhaust gases to generate steam. The steam is then passed through another set of turbine blades that convert heat energy into mechanical energy. The overall efficiencies of these turbines can approach 60 percent.

When the turbine design is used to collect energy from the wind or falling water, the efficiencies can be even higher. The principle is the same in converting the kinetic energy of the moving fluid into mechanical energy by the rotating turbine blades. An example of a turbine is the old windmills that dotted the countryside and were used to pump water. As was explained earlier, modern wind turbines can only achieve a maximum of around 60 percent efficiency, while water turbines can achieve efficiencies as high as 80 percent. Some kinetic energy must remain in the fluid so it can continue past the rotating turbine and not interfere with the incoming fluid.

Electric motors work the opposite of electric generators; however, both are identically constructed. In a generator, there is mechanical energy going into the system and electricity comes out. In a motor, electricity goes into the system and mechanical energy comes out. An electric motor works on the principle of the repulsion of like magnetic field and the attraction of unlike magnetic fields. Every child has been fascinated by the property of magnets to attract and repel each other. A motor is designed to use this property. It produces a continuous construction and collapse of electromagnetic fields in the stationary outer windings that continually attracts and repels the magnetic fields in the rotating windings. This simple design can be used both as a motor and a generator. If electricity is put into the system, the electromagnetic fields are created and mechanical energy comes out. If mechanical energy is put into the system, it drives the wires through a magnetic field and electricity comes out. Electric motors, as well as generators, can achieve efficiencies of around 90 to 95 percent.

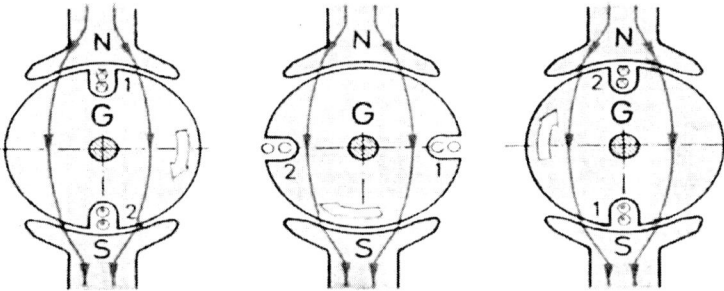

Fig. 5.9. As the wires (points 1 and 2) pass through the magnetic lines of force contributed by north and south magnetic poles N and S, current is generated in the wires (generator). If electricity is passed through the wires, a magnetic field is created, which is attracted and repelled by the magnetic field furnished by N and S, thereby producing a rotational motion (motor).

As fossil fuel use decreases over the next few decades, engineers, scientists, investors, and the general public will be looking at these processes to determine which will be used in the new energy paradigm. Efficiencies, availabilities, costs, and reliability will be some of the factors that determine what direction we will pursue. With the environmental problems the world is now facing, however, we must also consider the environmental consequences of any choice we make.

CHAPTER SIX

*Environmental Consequences of
Present and Future Energy Sources*

Any book that attempts to discuss what the future holds for world energy resources and possible future sources must also discuss the environmental consequences of the various types of fuels. As we have seen, within the next five to fifteen years, we must switch our dependence from fossil fuels to other sources of energy. Not only are we running out of the fossil fuels, but their environmental impact is affecting us daily and appears to have long-term consequences for our global civilization. As the earth becomes more populated, we must look for long-term sustainable energy sources that have minimal impacts on our global ecosystems. In this chapter, we will examine each of the pollutants that are now affecting our planet and try to determine what can be eliminated if we choose wisely when switching from our fossil fuel dependence.

There is still quite a lot of debate over the long-term effects of burning fossil fuels. Much of this discussion is fueled by purely economic interests, but some also show a valid concern for "good" science. The authors will make every effort to stick to the evidence that has been accumulated by the multitude of studies that have been made over the past decades.

Atmospheric Pollution

Humans, and other biological systems, live in a thin layer of air surrounding the earth called the biosphere. It extends from an altitude of ten kilometers (six miles) into the depths of the ocean. The normal concentration of gases in a dry atmosphere are primarily nitrogen (780,900 parts per million), oxygen (209,400 ppM), argon (9300 ppM), carbon dioxide (315 ppM), and trace elements of other gases including neon, helium, methane, krypton, nitrous oxide, hydrogen, xenon, nitrogen dioxide, and ozone. Pollution occurs when either a substance or thermal energy is expelled into the biosphere and has an adverse effect on biological organisms.

Air pollution is generally defined as those substances that are introduced into the air by humans and have a negative effect on the atmospheric environment. The air pollutants are in the form of gases, small particles of solids called particulates, or small droplets of suspended liquids called aerosols. The dispersion of air pollutants throughout the atmosphere is fueled by the earth's complex wind engine, which is produced by the earth's rotation and solar heating patterns. National boundaries mean nothing to this wind engine, and pollutants produced in one area often adversely affect the entire global biosphere.

The air pollutants that have an adverse affect on biological organisms are particulates, sulfur oxides, nitrogen oxides, hydrocarbons, carbon monoxide, and carbon dioxide. Particulates are extremely harmful to human respiratory systems, can further aggravate existing cardiovascular problems, and can do damage to the immune systems. Seven million tons of particulates are released into the atmosphere every year from industrial processing (primarily smelting) and electric power plants. Electrostatic precipitators are quite successful in removing larger particulates and are employed on most modern power plants. However small, sub-micron-sized particulates can escape the pollution control equipment and exist in the atmosphere for days before finally settling to ground or water.

Sulfur oxides, SO_2 and SO_3, are primary contributors to air pollution. When fossil fuels are burned, primarily high-sulfur coal or petroleum, the sulfur contained in the fuel is oxidized into sulfur oxides. High-sulfur coal contains as much as 6 percent sulfur by weight, and the burning of coal emits twenty million tons per year of sulfur oxides, about 98 percent of which is sulfur dioxide. There is a continuous recycling of sulfur into the environment from natural sources, primarily decaying organic matter (H_2S), and sea spray (sulfates, SO_4). These are less harmful forms of sulfur than the SO_2 emitted from burning fossil fuels. Studies on large populations show that SO_2 concentrations will increase illness and death rates, primarily on the elderly, the young, and those with existing respiratory ailments. Several major disasters have occurred in the past when meteorological conditions have produced very high concentrations of SO_2 and particulates. In 1948 in Donora, Pennsylvania, nineteen people died from increased SO_2 and particulate concentrations. In 1952 in London, four thousand people died from SO_2 concentrations that were seven times their normal level.

Also, SO_2, as it is further oxidized into SO_3, will combine with water to form H_2SO_4, a primary source of acid rain. The process of oxidizing SO_2

into sulfates occurs slowly, so the sulfur oxides released from one source can be dispersed over a much wider area before it falls to the ground as acid rain.

Nitrogen oxides (NO_x) are unique among pollutants, as they do not originate from the fuel itself but from the combustion process of burning fuels in our nitrogen atmosphere. The combustion of any fuel, therefore, will be a source of nitrogen oxides. At current levels, nitrogen oxides have a very minimal effect on human health; however, the sun will break the NO_2 into ozone (O_3) and NO. These compounds will combine with hydrocarbons from automobile exhaust and industrial plants to form very active organic radicals. The ozone and organic radicals are extremely harmful components of photochemical smog, causing eye irritation and respiratory problems. Also, nitrogen oxides will combine with water to form nitric acid (HNO_3) and fall to the ground as acid rain.

Scientists still do not have a complete understanding of the relationship between emissions of SO_2 and NO_x compounds and the creation of acid rain. The process, as described in this text, of converting these compounds into acid rain has been oversimplified and is really only part of the complete (not well-understood) process. Also, there are disagreements in the scientific community as to the long-term effects of acid rain on our biosphere.

Carbon monoxide and hydrocarbons are pollutants that primarily result from incomplete combustion in the firing chamber of the internal combustion engine. Since the 1970s, engineers have made advances in designing pollution-control devices that reduce these pollutants. In 1975 the average emissions of primary pollutants from an American automobile was hydrocarbons: 3.4 grams per mile, NOx: 3.1 grams per mile, and carbon monoxide: 34 gms per mile. In 1998 the average emissions of hydrocarbons had been reduced to 0.25 gms per mile, NO_x - 0.4 gms per mile, and carbon monoxide: 3.4 gms per mile. These dramatic increases in design efficiency have been somewhat offset by the greater number of cars and miles driven. Also, because of primary emphasis on economic growth in many developing countries, pollution control technology for those automobile industries has been ignored. As a result, in many large cities in the developing world with high levels of traffic, there are also very high levels of air pollution.

Mercury

Mercury is an element that has been mined for over two thousand years and has been known as a poison since that time. It is a controversial pollutant that occurs naturally in nature. Many in the anti-science block feel that since it does occur naturally in nature and is non-poisonous in some forms, it is not harmful and should not be classified as a dangerous pollutant. A lot of research found both online and in journals has shown it to be one of the more dangerous pollutants discharged in the burning of coal.

Mercury was first mined over two thousand years ago in the area of Almaden in Spain, where it was noted that most of the miners died. Because they were primarily slaves and prisoners, nobody was overly concerned and over the years, new uses for mercury were found. One of these was in the manufacture of felt hats, which produced mental illness in people who made the hats and led to the term "mad as a hatter" and to Mad Hatter's disease. The adverse effects of mercury were not studied until recently and are just now beginning to be understood.

Most people are familiar with liquid mercury, as it is used in home thermometers and thermostats. Mercury amalgams are made by mixing mercury and other powdered metal alloys and allowing the mixture to harden into an amalgams. They have been used as dental fillings since the beginning of the twentieth century and for other metallurgical purposes.

Mercury exists naturally in many forms in nature, as a liquid, metal, as a vapor, and in several minerals and compounds. The most dangerous form of mercury is monomethylmercury (MeHg), which is one hundred to one thousand times more toxic than elemental mercury. MeHg builds up in fish and when humans consume fish with high levels of the poison, they will accumulate the concentration into their bodies. In the 1950s, an accidental industrial discharge of MeHg into Minamata Bay in Japan led to the poisoning of local fishermen and their families. But eating fish is not the only way that it can enter the body. In 1972 a MeHg-fungicide treated grain was released in Iraq, which led to brain damage in several thousand people and the deaths of several hundred. MeHg seems to specifically target the human advanced central nervous system. Research indicates that the MeHg molecule is similar to some amino acids and once it enters the brain, it begins interfering with protein synthesis of the nerve cells. It is particularly harmful to developing brains because they are much more active (about three times) in amino acid transport.

Not all elemental mercury reaches the brain. It is often oxidized and trapped in the body's red blood cells. Some, however, is oxidized to divalent mercury that can reach the brain and produce Mad Hatter's disease (erethism). The American Academy of Pediatrics (AAP) in July 2001 issued a report asking parents and physicians to get rid of thermometers and blood pressure meters that contain mercury. The request was based on a well-designed study in the Faroe Islands that showed exposure to mercury (even at low levels) in children led to a decrease in memory, attention, and language skills.

About one-third of human contribution to the introduction of mercury into the environment comes from coal-burning power plants. Coal, depending on the vein from which it is mined, contains from 0.08 to 0.28 parts per million (ppm) of mercury. On a world scale, this means that coal plants contribute approximately five thousand tons of mercury to the environment each year. In the burning of coal, mercury is vaporized and later settles back to earth and is washed into lakes and streams. Microscopic plants and animals absorb the mercury and convert it to MeHg. Fish consume the smaller animals, which are then consumed by humans. Virtually every state in the nation now has restrictions on the amount of fish that should be eaten by consumers.

Because the effects of mercury are so insidious, there is much debate over how much mercury in the environment is actually harmful. Environmental and natural resource groups feel that we are well past the danger level, while economic groups feel that danger has not been well documented. As with so many environmental hazards, it is extremely difficult to measure direct relationships between cause and effect. For example, if there is a rise in mental disorders among the young, scientists have not determined absolutely if this is a result of exposure to mercury, genetic defect, other causes, or simply the result of improved reporting. Environmentalists claim there is sufficient data to show a relationship between cause and effect and are not willing to take a chance before it is too late. Groups with an economic interest feel it would be foolish to potentially harm the economy and their livelihood based on what they claim is insufficient data. Also, much of the mercury enters water ecosystems from purely natural sources. At this time, there is a tremendous amount of research investigating the relationship between mercury discharge from power plants and mental disorders.

Global Warming!

The effects of carbon dioxide (CO_2) on the earth are perhaps the most hotly debated in the scientific/industrial community and again, much of the debate is fueled by purely economic interests. With the large number of studies in different scientific disciplines all pointing in the same direction, however, the evidence for global warming as an accomplished fact is irrefutable. In the informed unbiased scientific community, there is simply no disagreement to these findings. In fact, the latest studies indicate that global warming is occurring at a much faster rate than was predicted several years ago.

Carbon dioxide is a gas that is given off by the burning of all fossil fuels. Fossil fuels, by definition, are carbon based and when any carbon-based fuel is burned, one of the products of combustion is carbon dioxide. CO_2 is harmless to humans and a necessary component of the photosynthesis process used by all plants. CO_2 is the fourth highest naturally occurring gas in the atmosphere, behind nitrogen, oxygen, and argon.

The problem is that CO_2 is also a greenhouse gas. That means the CO_2 molecule will absorb electromagnetic energy of certain microwave (infrared) frequencies. Virtually all matter is able to absorb energy at some frequency.

The process is similar to how an individual on a swing can absorb energy by another person pushing on it. As the swing goes back and forth, it can accept more energy from the individual pushing only if that individual pushes in tune with the frequency of the swing. If the person pushes at the wrong time, too often, or too slowly, it will interfere with the swinging motion. If the individual pushes each time the swing comes back and starts forward, the energy will be absorbed by the swing and will increase the amplitude of the swinging motion. In the same way, all molecules have vibrational properties that occur at certain frequencies and can absorb energy at those frequencies. One of the vibrational frequencies of the carbon dioxide molecule happens to be in the microwave (infrared) range, which allows it to absorb heat energy that is normally reflected back into space from the earth. There are many other gas molecules that vibrate at a frequency that will absorb infrared energy, e.g., water vapor, chlorofluorocarbons, nitrous oxide, ozone, and methane among them. Carbon dioxide, however, is by far the most abundant and therefore has the greatest potential for harm.

The earth receives electromagnetic energy from the sun at a variety of frequencies, from low-energy radio waves, through microwaves, including infrared, through the visible spectrum range, and through the ultraviolet range. The earth's protective atmosphere filters out most of this radiation except for electromagnetic windows, which allow certain frequencies ranges to penetrate to the earth. One of the largest of these is the window that allows the visible spectrum to penetrate the atmosphere, and much of the electromagnetic energy reaching the earth is in that frequency range. As the visible energy reaches the earth, it is absorbed by dirt, rocks, plants, water, people, and other creatures. Most of that energy is re-emitted by whatever absorbed it. The re-emitted energy is usually in a lower frequency range of the electromagnetic spectrum than was absorbed. Some of this infrared energy is emitted back into space and some is retained in the earth's atmosphere through the effect of the gases that absorb the infrared. In other words, a balance is made between the energy absorbed by the earth's atmosphere and the energy transmitted back into space. This balance keeps the earth's biosphere within a certain temperature range that has allowed for life to evolve and prosper. The gases that absorb the infrared energy emitted back into space are called greenhouse gases.

Greenhouse gases have played an important role in the evolution of life and our civilization. The average surface temperature of the earth is approximately thirty-three degrees higher than it would be without the greenhouse gases in our atmosphere. Atmospheric composition can also affect the surface temperature in other ways. Aerosols from volcanoes and water vapor from clouds can absorb and reflect sunlight before it reaches the earth, thereby cooling the earth's surface. A century ago, with the Industrial Revolution in full swing, concerns began to arise that with the large amount of carbon dioxide being pumped into our atmosphere, the planet may begin to heat up, a process called global warming. Over one hundred years ago, the Swedish chemist Svante Arrhenious stated that if the carbon dioxide concentration in the atmosphere were doubled, the average surface temperature would increase by five to six degrees Celsius.

For the last half-million years, the concentration of CO_2 has ranged from 200 to 290 ppm. From ACE 800 to 1800, it varied from 270 to 290 ppm. Since the beginning of the Industrial Revolution to the beginning of the twenty-first century the concentration has increased to 360 to 370 ppm. There is very strong evidence that we are seeing the effects of this increased concentration in terms of increased surface temperatures. This increase in surface temperature caused by humans pouring large amounts of greenhouse gases into the atmosphere is called anthropogenic greenhouse effect.

Although there is very little debate in the scientific community over the fact of anthropogenic greenhouse warming, there is much debate over the consequences of this warming. The primary problem in determining the consequences of this heating process is the complexity of the earth's biosphere. There are many processes the warming may trigger that will enhance or diminish the overall effects. These processes are known as feedback loops. Scientists simply are not sure what feedback loops the warming will trigger. One example is the warming will increase cloud formation due to increased evaporation of ocean and lake water. The increased number of clouds will reflect more incoming solar energy, which will offset the effects of greenhouse gases, diminishing the effects of global warming. Another is the oceans will warm, releasing some of their dissolved carbon dioxide into the atmosphere, which will enhance the effects of global warming. One feedback loop that is quite worrying to scientists is the effect the warming will have on ocean currents.

England and Northern Europe are warmed by an Atlantic current that brings warm water from the tropical areas into the North Atlantic. This conveyer belt of north/south Atlantic currents is powered by the salinity concentration and increased density of water as it moves north. With the northern ice caps beginning to melt and pour fresh water into the North Atlantic current, the density is decreasing and beginning to slow the current. What effect this may have on world and Northern European climates still remains to be seen. There is little evidence to support the exaggerated effects portrayed in the film *The Day After Tomorrow*, released by Twentieth Century Fox. It could, however, bring on a cooling of northern Europe, increased droughts in northern Africa, and an increase in Atlantic hurricanes.

Some scientists are now offering a range of possibilities that could occur as a result of global temperature changes. The best scenario is what we now see beginning to happen. That is a gradual increase in temperature over the next fifty to one hundred years. The increased CO_2 levels will bring greater and larger plant growth to the temperate zones with increased warming expanding the deserts. It will lead to the extinction of some plant, insect, and animal species because of the disappearance of their habitat. Weather patterns will become more intense because of the increased energy in the biosphere. Increased melting of the ice caps will add to sea levels with increased flooding of low areas. The worst-case scenarios are those of feedback loops being triggered, which lead to rapid climate changes and major impacts on world population and civilization.

As in the case of mercury poisoning, the effects of global warming are slow and insidious. Because the effects are so subtle, it is very difficult to get action from the government and businesses of an industrial country whose present economic security depends on contributing greenhouse gases to the atmosphere.

Over the past one hundred years, the technology of burning fossil fuels has increased tremendously. Internal combustion engines, turbines, external combustion engines have all about reached their theoretical maximum in efficiency. There are techniques for scrubbing many of the pollutants from the exhaust of fossil fuel power plants. There is now pre-combustion heating processes that look to be very promising in removing some of the mercury and other poisons from coal. The one product of combustion that cannot be removed and is a part of all fossil fuel combustion is carbon dioxide. Advocates for the use of coal as the primary fuel for future electrical generation are now discussing carbon sequestration as a method for disposing of the carbon dioxide from combustion. This simply means burying carbon dioxide in large non-atmospheric sinks (usually underground caverns or old wells). This solution is very short term and, as we will see, takes up storage space that will be needed in the coming hydrogen economy. Also, all of these efforts in combustion technology and carbon sequestration add a large cost to coal-fired electrical generation. Many companies are fighting even basic improvements. If all were incorporated into present plants, the costs for electrical energy would be much higher than moving toward the solar technologies.

In the United States, 99 percent of our transportation and 75 percent of primary electrical generated power comes from fossil fuels. The consequences of global warming are now becoming more apparent and will probably be the straw that finally breaks the back of our dependence on fossil fuels. We simply cannot continue this destructive experiment on the earth's biosphere by dumping tons of CO_2 into the atmosphere daily.

CHAPTER SEVEN

Where Do We Go?
How Do We Get There?

A primary argument against transition to a sustainable economy is the sheer volume of our national energy needs. As we will see, if we work from the bottom up, using local resources and imagination, it is an attainable goal. In this chapter, we will briefly discuss the process for arriving at a sustainable economy. The numbers that justify the process, for the readers who are interested, can be found in Appendix C.

In this country and over much of the world, we have become very dependent on federal and state governments or large businesses to solve our problems for us. As we have seen, with the rise of interest groups, the government has become quite ineffectual in dealing with long-term problems if the solutions offend any sector of our population, no matter how shortsighted that sector is. Business is primarily driven by market forces, which can also be quite shortsighted. This is not to say that business and government will not play a part, but the driving force for the transition must be at the local level. The process of attaining a sustainable energy economy must begin with a bottoms-up integrated approach.

The United States uses about ten thousand billion kwh of energy per year in our electrical generation industry and about the same amount in our transportation industry. (This total amount comes to around 80 to 90 exojoules per year, which are the units used in appendix C, 1 exojoule = 10^{18} joules = 1.72×10^8 barrels of oil.) Any path we begin will have to account for that total amount of energy. It is very easy to say we should switch to a hydrogen economy, natural gas, coal, or renewable technologies, but unless that switch accounts for our environmental needs, our total energy demand, and the process for getting there, we simply have words being tossed around. As was mentioned above, the path we will look at is twofold. It must start at the local level and must be an integrated approach. No single source of energy will supply all of our needs, and the various energy resources available on a local level are best recognized by informed local residents.

Another major advantage of generating energy at the local levels is the increased efficiencies. If we light an incandescent bulb from electricity that is generated in a coal-fired power plant hundreds of miles away, it is a 1 percent efficient operation. That is if one hundred pounds of coal is consumed with all the environmental effects, only one pound of the coal goes into lighting that light. The rest is lost in heating the rivers the plant is on, heating the air the lines run through, and heating the area around the light bulb. If we generate locally, we do not have the line loss and the 70 percent of the energy that normally goes into heating the rivers can be used locally to heat buildings, generate secondary power, and also be used in manufacturing processes.

Integrated Approach

A six-step integrated approach should be implemented on local, state, and national levels. The process should start by identifying possible sources of energy and storage methods at the local level through communities, co-ops, public utilities, and other financially or ideologically interested groups. Local, state, and federal laws need to be passed that will ease the transition into a sustainable economy. Many of the laws associated with energy were passed at times of different energy scenarios. They are quite outdated for present needs. We will examine some of these laws later in the chapter.

The six-step approach that every community should consider must include:

1. Raising public awareness to obtain the needed support.
2. Implementing conservation measures.
3. Considering various energy storage methods.
4. Identifying local renewable energy sources.
5. Implementing distributive generation using the sources identified.
6. Beginning the process of converting to a hydrogen economy.

These steps are integrated, and initially all should be considered at the same time.

There is no set order, and each should be continuously reconsidered as new information and technologies become available.

Raising Public Awareness to Obtain Needed Support

"Not in my backyard" has become the mantra of many modern Americans. Sometimes this is a valid sentiment but as a blanket statement, it can lead to stagnation in local communities. The authors believe, and the actions taken by some communities confirm, that this does not have to be the situation. If community leaders, the press, and local businesses get behind an idea and include the general public from the start, local support can be generated. Generating this support should be started immediately when a community decides to take action, and the process should be continued throughout the whole project.

Raising public awareness and changing how communities view the lifestyle changes needed to implement a local sustainable energy economy is perhaps the most important of the six steps. Nothing will be accomplished without it. The process of changing norms and how the public views social norms is discussed more fully in chapter eight.

Implementing Conservation Measures

In the 1970s, during the OPEC oil embargo, the single most effective measure taken to reduce dependence on OPEC oil by individuals and the government was the increased energy conservation. *Scientific American* reported that more energy was saved from the mid-seventies to the mid-eighties through conservation than was produced by any new oil fields or coalmines. This was accomplished while our gross domestic product rose 20 percent. Individuals began purchasing more efficient products. Local communities instituted energy fairs, where manufacturers could demonstrate their energy-saving products and speakers could discuss methods of saving energy. Local co-ops and public utilities started lending programs, where customers could borrow money to purchase the energy-saving products and pay for them on their monthly bills. They conducted energy audits of buildings that showed where insulation and state-of-the-art windows and doors were needed. The federal government passed regulatory laws mandating minimum miles-per-gallon standards on fleets, and laws were passed to allow electricity generated with renewable sources to tie into the grid system. Some states and the federal government offered tax credits for the installation of energy storage and renewable energy systems.

Over the past thirty years, many of the measures taken in the seventies and early eighties have become part of our standard operating procedures with increased efficiencies in the construction, manufacturing, and production areas of the economy. We need to build on what was started in the seventies and examine our homes, businesses, and transportation to determine methods of saving energy. One of the added benefits of the energy conservation drive during the seventies was that everyone, from housewives/husbands to businessmen/women, seemed to get involved. If we again work from a local level, we will again see this involvement.

Energy conservation can be achieved through two different measures: either by lifestyle changes or technological fixes. Sometimes a lifestyle change is needed to adopt the technological fix. A good example is in the transportation industry. Hybrid cars have been on the market for several years but beyond a few individuals have not gained wide acceptance. They are not as roomy and do not have quite the performance as most completely internal combustion engine vehicles. It would take only a minor lifestyle change to switch from the roomy SUV, van, or standard car to the smaller hybrid. Most are unwilling to do this and as a result, hybrids are not being accepted by a large percent of the population. The latest increase in gas prices to around $3.00 a gallon has demonstrated to some the need for hybrids and they are becoming more popular—in fact, at this time, there is a waiting list to purchase one. But if gas prices were to go down, the waiting lists would disappear. Petroleitus has a very short memory. This scenario will continue over the next few years. Projections indicate that within ten years, as prices continually increase, 90 percent of the cars sold will be hybrids. If every American were to purchase a hybrid vehicle, we could dramatically clean up pollution in large cities and cut our national transportation energy consumption. We could completely eliminate the Middle East as a source for purchasing oil. Mass transit is another example where a lifestyle change could produce large savings in our transportation energy use. Most Americans are, however, addicted to the use of personal vehicles.

It is difficult to overestimate the energy savings that could be accomplished if the U.S. population wholeheartedly endorsed conservation measures. David Pimetel, one of the world's leading energy researchers, believes that with super insulation, green architecture, more efficient motors, energy-saving vehicles, more efficient lighting, efficient washing machines, and other conservation methods, the U.S. could cut its energy consumption by 33 percent. He feels the major roadblock to conservation measures is the low cost of energy brought on by subsidies to the energy industry. Pimetel states that every American is paying about $410 a year in taxes for subsidies to the energy industries that keep the price of energy artificially low. If these subsidies were removed and the price of gasoline, heating fuel, and other energy-related products was allowed to reach its true cost, conservation measures would be implemented that would produce a net savings in cost and energy consumption.

Energy Storage

After energy conservation, energy storage is the single most important step needed for a massive energy savings and the introduction into a sustainable economy. Without developing and using energy storage there is simply no possibility of attaining a sustainable energy economy. A primary criticism, and valid one, of switching to solar energy technologies is their intermittent nature. The sun does not always shine and the wind does not always blow. Calculations show that without some form of energy storage, there will be no decrease in the size and number of fossil fuel power plants required. Also, because all heat engines run most efficiently at high continuous outputs, the fossil fuel power plants will run even more inefficiently if they must be started and stopped continuously as they are needed (no sun, no wind). These inefficiencies will contribute even more to the environmental problems associated with fossil fuel.

There are several storage processes that could be implemented that would help solve this problem. The first could be on-site thermal storage for business and residential buildings. When the wind is blowing, the sun shining, or the grid power demand is low, a signal can be sent to buildings that would turn on the heating or cooling system. The excess power from the grid would be stored in the form of thermal storage at building sites. Renewable energy sources would be much more attractive on a large scale if all new buildings incorporated some form of thermal storage and older buildings were retrofitted with thermal storage.

A second method would be to use hydrogen storage coupled with wind generators and photo voltaic cells that are completely dedicated to hydrogen production. During times of wind or sun, the systems will be producing and storing hydrogen. In times of no wind or sun, the hydrogen can be used for generating electricity, either through fuel cells or turbines. This process would also allow for direct current generation of electricity to produce the hydrogen. Generating DC current allows wind energy systems to operate over a larger range of rotational speeds (rpm range) at greater efficiencies. The hydrogen would be stored in pressurized tanks, hydrides, pipelines, or large underground caverns. As hydrogen fueled cars become more common, the stored hydrogen could also be used for refueling.

There are several other methods for storing energy that are more site specific, e.g. pumped water storage for regions that have elevated areas that could store the pumped water for later electrical generation. Also, underground compressed air or hydrogen storage could be used in areas that have underground caverns and depleted natural gas wells, just as natural gas is now stored.

Local Energy Sources

The two most obvious sustainable energy sources are wind and solar power. These should be harnessed wherever possible. Over the past several years, there has been rapid growth in large wind energy systems. It is now the fastest-growing energy source in the world. It is extremely competitive in cost and could be even more competitive if all subsidies were removed from fossil fuels. There should also be a market for smaller systems that could be installed by individuals and small farms if local laws, utilities, and coops were more installation friendly. This has been a problem since the seventies in many areas and should be examined by state and local governments.

There are areas, however, where neither wind nor solar is practical and other sources should be considered. An example is northern Minnesota, where days are short in the winter and there is not much wind or sunshine. The area does have quite a lot of biomass in the form of crop residues and trees. Studies could and should be made to determine how much energy can be extracted from these biomass sources in a sustainable manner. The biomasses can be used to fuel steam turbines for distributed power generation or more directly in a community hot water steam heating system.

Several years ago, the Pollution Control Agency in Minnesota was confronted with a problem of flax chive residue completely overwhelming some northern Minnesota landfills. Local farmers raised flax as a cash crop and would extract the fiber and the oil from the plant. The husk (chives) remained with no practical use. Since chives are primarily carbon, the farmers did not want it in their land since it took quite a lot of time and expensive nitrogen fertilizer to break down. They initially disposed of it by depositing into local landfills. The farmers and landfill operators went to the PCA looking for a possible solution. The PCA conducted some experiments and determined the chives were an excellent clean fuel for generating steam when burned in suspension, a technology readily available. Carbon dioxide was released during the combustion process but would be removed from the atmosphere the following year, as new flax plants were grown with no net atmospheric gain in CO_2 (carbon neutral).

The PCA held meetings in an attempt to interest local manufacturers and school districts into using the flax chives as a heating fuel. The technology for burning in suspension is well developed and not terribly expensive to implement. The fuel was free, and the farmers promised to deliver anywhere within a thirty-mile range. There were no takers. Energy was cheap at the time, and the participants did not want to be bothered with implementing the technology. Nothing came of the meetings but demonstrated what *could* be done at local levels if the interest and need is there.

There are many forms of clean sustainable energy that could be harnessed if local creativity were put to use. Geothermal energy, tidal energy, ocean temperature differentials, wave energy, solar ponds, crop residue, and animal waste could all be used to generate energy in local communities.

In areas of northern Europe, some livestock yards and farmers are using animal waste in digesters for generating methane that is stored in large balloon/dome-like structures. During peak electrical loads, the methane is burned to power steam turbines, producing electricity and supplying extra income for the stockyards. The remaining slurry is sold as an extremely nitrogen rich fertilizer. Also, in northern Europe, they are developing large community-sized waste wood steam generators for community heating systems. The authors use free edgings (waste wood) from a local sawmill to heat their swimming pool (thermal mass) in the winter, which then heats their house.

Distributive Generation

Distributive generation is defined as the generation of energy at or near the location where the energy will be used, with smaller generating plants, usually in less than ten megawatt generating systems. It is certainly not a new concept, and virtually all co-ops and utilities got their initial start in distributive generation. Because of the improvement in efficiencies using economy of scale, however, large centralized power-generating plants became the norm. With the tremendous increase in efficiencies in smaller systems, using new technologies, primarily the computer, turbine, and fuel cell industries, distributive generation is again looking to be the energy paradigm of the future. There are many advantages to distributive generation that appear to make it the obvious choice in this future scenario. The business and governmental forces that created our present large-scale energy system, however, have a major influence over perceptions and political power that make a transition difficult.

A nationwide system of distributive generation would be much less susceptible to blackouts or terrorist attacks on our national grid. It would allow for co-generation where the waste heat from electrical generation could be used for various purposes in the local community, e.g., heating, manufacturing, etc. Distributive generation can be well matched to local load requirements, especially if businesses decide to install their own generators and also would increase jobs at the local level. Most important, distributive generation would allow for local resources to be used in the energy generation process.

As an example, let us return to the northern Minnesota community that has an excess of biomass that can be used as a fuel. Basically, the fuel is free or nearly free with the cost coming from the gathering of the biomass. Once the biomass has been gathered, it can be combusted in a high-temperature steam generator either through directly burning the biomass or, more efficiently, by first grinding the mass and burning in suspension. Once the steam is generated, it can be used to produce electricity in a steam turbine powered electrical generator. One of the most efficient is called a Rankin cycle microturbine. Once the steam has been used to generate electricity, it can be transferred to other areas where low-temperature steam can be used for manufacturing or heating purposes. This process gives an overall efficiency of nearly 70 percent.

Some areas of the country may have an energy source that would not provide for the high temperatures needed to generate steam, e.g., geothermal or ocean temperature differentials. Other working fluids, in place of water, that vaporize at lower temperatures could be used in the Rankin cycle microturbine.

Converting to a Hydrogen Economy

Jeremy Rifkin in *The Hydrogen Economy* discusses how civilizations rise and fall around energy uses and needs. He sees a whole new gentler, more inclusive, and more empowering world civilization arising with the hydrogen economy and fuel cell technology. Rifkin envisions a worldwide web of hydrogen sharing based on the worldwide web of information sharing that has developed along with the computer. It is a very compelling and idealistic vision of an energy future that he feels is very attainable.

Although the hydrogen economy envisioned by Rifkin is something that should be pursued, it certainly will not happen overnight. We need a shift in paradigm and an infrastructure to accommodate the hydrogen economy. We must look to the history of the beginning of the petroleum age and the introduction of electricity into the twentieth century to get an idea of how paradigms change and how infrastructures grow. It was almost seventy-five years between the development of the electrical generator and the mass electrification of the United States. It took more than fifty years and World War I between the development of the internal combustion engine and the complete evolution of petroleitus. Paradigms gradually changed, and the infrastructure gradually developed that led to the electrical and petroleum civilization we see today. If we begin with the six steps outlined in the first part of this chapter, there will be a gradual evolution into a hydrogen economy. Also, because technologies are introduced into society at a faster rate in the present than the past, we could possibly arrive at a full-scale hydrogen economy within thirty to forty years.

The conversion must start at the local level with communities, co-ops, utilities, and private entrepreneurs producing, storing, and selling hydrogen for local use. Gradually local production will evolve into a national network of hydrogen sharing, just as the electrical grid and natural gas pipelines have evolved. The hydrogen network will be able to build on the natural gas distribution pipelines that are already in place.

Several large automobile and petroleum companies have begun the transition into hydrogen by designing hydrogen cars based on fuel cell technology. They use hydrogen produced by stripping hydrogen atoms from fossil fuels, primarily natural gas. This is not the most feasible nor cleanest way to get to a hydrogen economy. By stripping hydrogen atoms from fossil fuels, some carbon atoms from the fossil fuel remain in the hydrogen that poison the fuel cells. Eliminating the poisoning problem in fuel cells is a difficult and expensive technological fix that is not required if the hydrogen were produced from renewable energy sources. Second, the stripping process releases carbon dioxide into the atmosphere. One of the primary reasons for eliminating fossil fuels is to curb global warming due to carbon dioxide in the atmosphere. Advocates of using fossil fuels to produce hydrogen want to sequester the carbon dioxide by placing it into old wells or underground caverns, again a very short sighted approach. The old wells and caverns could and should be used for storing hydrogen.

Role of the Governments in Transition to a Sustainable Economy

As was mentioned earlier in the chapter, the federal government is so strongly influenced by well-funded special interest groups that it appears to be almost ineffectual in dealing with major problems on a national level. There are many bright and farsighted members of Congress who have hardworking and insightful staff doing their research. The economic and political power of the fossil fuel interest groups, however, simply drowns out these voices. Given the environmental gains, the economic advantages, and increased security, there is simply no compelling reason why our national economy should not be converted to a sustainable hydrogen energy base. But in spite of these benefits, a very large percent of the national and international funding is going into fossil fuel research, applications, and subsidies. It appears the best we can hope for at the national level is that it does not hinder the beginning developments of a sustainable economy at state and local levels.

Some of the national laws passed in the last fifteen years do allow for distributed generation. Now these laws must be implemented and enforced at the state and local level. Many local co-ops and utilities have practices that make it almost impossible and/or economically unfeasible for small sustainable energy systems to connect into the grid. Many of these co-ops/utilities use safety questions as the primary hindrance for preventing the connections. Safety certainly should be a major concern. The states should implement procedures that address these concerns and possibly supply technical assistance to local co-ops/utilities to ensure safe interconnection. Once the safety issues are met, the sustainable energy systems should be allowed immediate connection into the grid. This rarely happens.

There are two valid technical concerns that should be investigated as more distributed/sustainable energy systems interconnect with the grid. The first concern will be to address the reliability of maintaining a clean sixty-cycle wave transmitted over the grid. In most wind systems and many distributed generation systems, the sixty-cycle grid current forces the generation of sixty-cycle current in the local generator/alternator. We do not yet know the maximum generating potential that can be controlled by the sixty-cycle grid before it is overwhelmed by the integrated local generators and no longer generates at sixty-cycles or the wave form becomes distorted. At this time, there is certainly no problem, but if distributed/sustainable energy reaches its full potential, the condition of the transmitted wave is something that has to be monitored and addressed.

The second technical concern involves that of hydrogen storage. The hydrogen molecule is very small and under pressure can seep into the crystalline structure of many metals. This can cause the metals to become quite brittle and crack. Will the structural integrity of the natural gas lines be compromised when the hydrogen web becomes a reality and the hydrogen is transmitted through the gas lines? Again, this is a possible problem that should be monitored and addressed.

Other concerns will arise as we move into a sustainable energy future. The benefits, however, far outweigh any possible problems that may appear. Many of the Northern European countries have made the commitment for the conversion into a sustainable economy. Their national economies are expanding at the same time their dependence on fossil fuels is decreasing. The United States must make this same commitment. Using the six-step integrated approach outlined in this chapter, we can and should convert to a sustainable energy economy.

Chapter Eight

Social Change and Energy

This entire book tells us the history of energy so we may plan a sustainable energy future. In order to plan for such a future, we must understand a little bit about social behavior and social change. A large part of social behavior is predominantly due to the culture in which we are born. Culture is simply the way groups do things. It involves material things, such as artwork and books. It also includes nonmaterial aspects of who we are, such as levels of power, language (both verbal and nonverbal), ideas, values, beliefs, and attitudes. Social norms, the ways we are expected to behave, are also part of our culture. We learn the social norms of our culture through socialization. Socialization is a lifelong process and occurs when our families, peers, media, education, employment, government, and even daycare tell us how to behave and what to believe. Whenever we come into contact with a group, we are socialized into its norms. Simply put, socialization is how others teach us what is "right" and "wrong" according to our culture. It is also how we teach others such as children and friends. The social norms learned in our culture are powerful. Many of us have difficulty even identifying these norms much less going against them.

Social norms are one of the biggest reasons we use a certain type of energy. Alternatives are difficult for us to consider because we are so used to what we already do. When social norms are widespread and there is enough demand, suppliers can keep energy affordable. One of the reasons people have not used many of the alternative energy sources is because they are not the norm. This creates some barriers to using alternatives because people fear the unknown, people dislike being labeled "different," and because some sources are difficult to get because they are not available on a wide scale. This also means alternative sources may be expensive. Furthermore, energy is a political tool and many politicians attempt to persuade and negotiate for specific types of energy sources, ones that benefit their states, their companies, their donors, and even their families and friends. Unfortunately, while supporting one source of energy, people often criticize other sources. Given these social and physical barriers, it seems social change around energy is going to be a long battle. But not necessarily. Look at the history of energy use outlined in this book. It is all about social change. Although we have social norms that most of us follow, we are also unique and interact in ways that change the world.

What Is Social Change?

Social change is the transformation of culture over time. In order for something to be considered an actual change, it must have changed a pattern or patterns of society and the change must last a period of time. Social change can be summed up as a long-term shift of social norms. Many energy changes discussed in this book are clear examples of this type of social change. In fact, this book demonstrates that social change around energy has happened many times throughout our history.

Wind energy is a perfect example of this because wind power now supplies the world with ten times the amount of energy it did ten years ago and is the fastest-growing source of energy in the world. It has been an interesting process. The wind is free. Why don't more people use it? The answer lies in the availability and affordability of the machinery associated with it. Simply put, if it were mass produced, it would become affordable for most families. Also, if it were subsidized in a way similar to fossil fuel, it would be more affordable. Another problem is that people may think it would be a bother since our current structure is so user friendly. Of course, the rising costs of energy demonstrate that although it is often a user-friendly system, it is not always affordable to everyone.

It is clear that some people are making the change to sustainable sources of energy whether driven by materialistic beliefs (saving and making money) and/or moral environmental beliefs (sustainable energy maintains our quality of life while protecting the same quality of life for future generations). It's obvious there have been many shifts in the types of energy used and the way people use energy. Much of this has been intentional and for those people who are interested in making social change to a more sustainable energy source, a discussion of how social change occurs and some examples concerning energy follow.

How Does Social Change Occur?

Social change occurs in a variety of ways, but it's always because people make change. Although there is social change that is unintended, we will focus on social change that is intended and deals with energy. In order to understand what we can about social change, it will be discussed as happening on various levels ranging from the individual, to community and national with a brief discussion of the global scale.

The causes of social change are complex and although this list is not all inclusive, causes can really be summed up as falling into two categories: human caused social change (e.g., culture, social norms, technology, materialism, social movements, social reform, laws, policy, revolution, and war) and nonhuman caused social change (e.g., earth quakes, storms, weather patterns, tsunamis, and hurricanes). Although this seems simple, it is sometimes difficult to determine cause. For example, the majority of scientists agree that most of our global warming is caused by human uses of fossil fuels and other global warming gases that have been increasing since the Industrial Revolution. A rise in global temperature may seem to be a nonhuman-caused global pattern, but it is linked to human behavior. This causes social change because groups have to find ways to deal with global warming.

This book promotes a six-step approach to intended social change concerning energy. We'll explore each of the six steps as they relate to specific levels of change: individual, community, and national. It's important to remember these are merely suggestions. One has to take responsibility and be creative and motivated. Most importantly, remember that planning and implementing are very different. Being flexible during all stages is vital to being successful. As this book is a testament, you can see how successful people have been even when others have told them what they want to do is impossible.

Change at the Individual Level

The first step in the six-step approach is raising public awareness. Although this section discusses individual change, it is important to note this level of change often creates change in others. While the individual informs themselves, they often share that information with the people in their life. People interact and impact one another, so when an individual changes, it may impact those people around them into making change. Charismatic leaders such as Eleanor Roosevelt, Jesse Jackson, J.F. Kennedy, and Martin Luther King Jr. all inspired others to change. For many of us, there are people in our daily lives who inspire us to change, including friends, family, and even neighbors.

The second aspect of the six-step approach is implementing conservative measures. One way to begin this process is to complete a self-assessment of your energy lifestyle or, if you have a family, of you and your family's energy lifestyle. How much and what type of energy do you use in your household? In your daily life (this includes work and recreation)? In your automobile, if you have one? Do you fly to conferences or to see relatives? If yes, how often? All of these things take energy and as an individual, you can look at alternatives to each of them. Based on your assessment, you can make changes specific to your lifestyle. For example, you may improve the insulation of your home, purchase hybrid vehicles when you have the opportunity, and drive less while walking and biking more. You could purchase sustainable energy for your household, such as solar cells, small wind generators, and fuel cells. Turn off microwaves, computers, and TVs, and don't leave your cell phone plugged in through the night. Many of the ideas concerning individual change were mentioned previously in this book. Ideas from Pimetel, such as super insulation, green architecture, more efficient motors, energy-saving vehicles, more efficient lighting, efficient washing machines, and other conservation methods, were linked to potentially cutting the U.S. energy consumption by 33 percent.

The third step in the process is considering various energy storage methods. Remember, this is the key to our energy problem. Is it possible for you to install conserving storage techniques in your home? Some ideas put forth in this book include new buildings incorporating some form of thermal storage and older buildings being retrofitted with thermal storage.

Identifying local renewable energy sources is the fourth step. This may include sun, wind, hydro, and even wood if you have a wood lot that replaces trees at a higher rate than usage. If there are none, then create one (flax, water, or wood lot). Remember, the two most obvious sustainable energy sources are wind and solar power. These should be harnessed wherever possible.

The fifth step is implementing distributive generation using the sources identified. Distributive generation is defined as the generation of energy at or near the location where the energy will be used, with smaller generating plants, usually in less than ten megawatt generating systems. Once you have identified a sustainable energy source, you should work to use it.

Finally, begin the process of converting to a hydrogen economy. How does your energy lifestyle link in to the overall energy system? Making changes should take into account the potential for a hydrogen future and being mindful of how you would connect into a hydrogen system.

Before moving on to the next section to discuss energy and social change at the community level, there needs to be a discussion about information. Information is often the key to social change. We rely heavily upon media information, and it is the rare person who investigates further. The reality with much of the media is that it is one-sided and often for the purpose of entertaining. Because of this, people should inform themselves.

When making social change, be sure the research you believe in is "good" science. Here are some hints:

1. Make sure the source is reliable. Who did the research? Was it someone who is being paid by the organization they are supporting? For example, if a scientist is employed by an oil company and endorses a product made by the company, that is highly suspect and the information should be considered less valid and reliable than scientists who are independent. It doesn't necessarily mean the information should be disregarded, but further investigating needs to be done on the topic.
2. Use more than one source of information and rely on peer-reviewed journal articles for information. Many people do not know that peer-reviewed journal articles are only published after experts in the field read the article. This is important because the experts have in-depth knowledge of the topic being covered. They often know when a study isn't done correctly or the data is faulty. The result is they do not support publication unless it is up to their standards. You can usually find journals on the web, in local libraries, and/or by asking people in the field about where you can find the information.
3. Make sure your information is complete. Is all the information available so you can clearly understand the research? For example, we often hear a newscaster make a comment using a statistic like: "Drug use is up 15 percent." Since you do not have all the information about this study, you are left to make many assumptions. Drug use is up with whom? Compared to when? Where was the study done, urban or rural? With what age group? Who did the actual study? How was the information gathered? If you do not have the answers to these questions, you should be careful with the information even if it is something that makes sense to you.
4. Make sure the language is clear. Is the study reporting with misleading words? For example, we often hear the words "clean coal." There's really no such thing as clean coal. Maybe cleaner coal. This is totally misleading and may cause people to support an energy plan that would have harmful consequences. Clean coal does not mean the coal is completely clean and safe. People use these words knowing they are misleading others into feeling comfortable. It is important that we consumers be aware of these tricky terms and inform ourselves enough to know their meanings. There's nothing wrong with saying, "Let's use cleaner coal until we get a better alternative that is more sustainable." People are not dumb, but we can be misled.

5. Be aware of social norms that create issues around energy such as pollution as a result of the type of energy we use and wasting energy. The example of clean coal is a good one for this point as well. If we believe coal is clean, then we will not take steps to protect ourselves and our children from the harmful effects. We have such great quality of life here in the United States. We need to be careful to develop social norms that protect. We also must emphasize that change may mean we work a little harder and pay a little more in the beginning but usually over the long run, we benefit.
6. Calculate more than merely finances when you calculate costs and benefits. For example, using sustainable sources not only means they are renewable but they decrease pollution in our atmosphere, therefore bringing down healthcare costs. What are the costs of pollution? What are the costs to communities for the clean up of toxic spills?

As mentioned at the beginning of this section, individuals making change can often create change in others. It is important to be a good role model for conservation and moves toward sustainable energy. With this in mind, it is prudent to note that the authors of this text live in a super-insulated home with passive solar (large windows facing South). We drive a Honda Insight hybrid commuting vehicle, recycle, and often purchase clothing at used clothing stores. Although these changes have been relatively easy, we have more changes we would like to make that are specific to our lifestyle. Many people and the media have been to our home to learn about it, and we hope people come away with more information to help them make sustainable energy changes in their lifestyles.

Change at the Community Level

The first step in the process is raising public awareness to obtain the needed support for change. Without community support, social change can be difficult. To accomplish the support, include as many community organizations (religious, civil, and business), the local media, the city (or township) government, and county government in the planning phases. This step will involve organizing and planning. You may want to develop a presentation that informs community members about the strength of developing sustainable energy at the community level. This presentation should be specific to the community but could involve some basic information about why sustainable energy is better than non-renewable energy sources. For example, sustainable energy pollutes less, is more secure related to terrorism, creates healthy jobs, and protects the livelihood of future generations.

Concerning the second step, implementing conservative measures, help your community develop a group of people who can complete an energy assessment for your area. If there is nobody in the community qualified to complete an assessment, there are many organizations that will complete one for a fee. The energy assessment plan should focus on ways the community can conserve energy but should also help with steps three (developing appropriate energy storage methods), four (identifying local renewable energy sources), five (implementing distributive generation or using those local energy sources for community energy), and six (ways your community can prepare for a hydrogen economy). Do not hesitate to consider how your plan may increase jobs or offer better jobs in your community. Use this information to develop a sustainable energy plan that includes clear goals, objectives, and a timeline.

One good example of this level of change is something that is happening in Minnesota. The Minnesota Project, the University of Minnesota's Regional Sustainable Development Partnerships, the Rural and Metro County Energy Task Forces, and the Resource Conservation and Development Council have all joined together to address sustainable community energy needs. The project was designed to help the community develop plans and have access to sustainable energies. According to the manual *Designing a Clean Energy Future: A Resource Manual Developed for the Clean Energy Resource Teams* (CERTS) by the Minnesota Project, University of Minnesota's Regional Sustainable Development Partnerships, and the Minnesota Department of Commerce (2003), "The outcome of the project will be a comprehensive and strategic renewable energy plan and vision for each region that reflects a mix of energy sources, such as biomass, solar, hydrogen, and wind" (viii). The focus of the project is developing "community energy...based on electricity generation that is located in or near the building, facility, or community where it is used." This is also known as distributed generation, which was mentioned previously and is renewable energy like wind, biomass, hydropower, solar, and some fossil fuels like diesel and natural gas (often used as transitions to something more sustainable).

According to the CERTS manual, there are eight steps to community energy process. These include the following: "1. Agree on common goals. 2. Raise community awareness. 3. Form a steering committee . 4. Gather and examine information and data. 5. Start with efficiency upgrades and conservation. 6. Develop an action plan . 7. Turn the plan into action. 8. Evaluate and build on success." Organization is the key to success when making change. Following a simple plan with goals and objectives allows groups to see when successes are reached. This eight-step approach is a workable one and linked to the success of community projects.

One such successful case study identified in the CERTS manual is the "Phillips Community Energy Cooperative: Consumers Control of Energy

Use." This community developed an urban energy cooperative, allowing the community to have more control over energy use and "to link conservation programs with under-served populations." One goal of this community project is to develop a renewable biomass combined heat and power facility "that would provide district heating and cooling to Phillips neighborhood business and residences." This project would develop a sustainable energy source to the neighborhood while providing jobs at the same time.

The Phillips Community Energy Cooperative (PCEC) has over two thousand five hundred members and has implemented programs such as "energy captain," where leaders organize their street or block to promote energy efficiency and encourage people to become members of the PCEC. According the CERTS manual, residents "can join the cooperative for $1 and then receive a free 'energy efficiency' kit, which includes: two compact fluorescent light bulbs, one low-flow shower head, and a 5-pack interior window insulation kit." The PCEC also has air conditioning trade-ins, energy upgrades, and a refrigerator program. All of these programs have been successful and have resulted in residents of the neighborhood saving money and transitioning to more sustainable energy sources.

The CERTS manual can be found at www.cleanenergyresourceteams.org/publications and also has a section titled "Helpful Resources for Communities," which lists organizations, their missions, and their websites. These sites are helpful in countless ways to groups who are organizing their communities to make sustainable energy changes. Getting on the web and learning about communities who have already made successful change is a good way to help focus your group.

There are some other hints related to community change. If you are having trouble financially making a change, get with other people and pool your funds, write for a grant, and/or ask a business to sponsor you. You may have to do more than one of these. Community forums can take place at libraries, restaurants, local schools, and clubs. Support environmentally friendly businesses by holding environmental open-houses at the business. Give your community a criteria for sustainability specific to your community. Accompany this with a list of goals and as each goal is reached, have a celebration. Have awards given to local businesses and individuals a few times throughout the year. Show businesses how to be environmentally friendly so your organization will support them. Come together as a community that is informed, organized, and demanding about its energy use. As communities inform one another, make decisions about supporting energy plans on the national level. Use your community momentum to send messages to the next level of change. Be vocal, demanding, and choosy about who and what you are willing to support.

Change at the National Level

Social change at this level is often a power play among individuals, organizations, corporations, political parties, and political action committees (PACs). PACs have much power in the field of politics, while political parties have less power than in the past. PACs are politically savvy groups that are hired to push somebody's cause, such as cotton growers, beef ranchers, gun manufactures, environmentalists, etc.

On one hand, it becomes important to know which PACs support sustainable energy policy and which ones do not because of the type of power they have influencing politics. Maintaining membership and support to PACs that promote sustainable energy is important. Maintaining political contact in other ways, however, is also important. Informed voting, letter campaigns, and news conferences are some additional ways to demonstrate your strength as a community. Make sure your community voice is heard by your representatives.

Demanding governmental support for sustainable energy social change is the important part of this level. Electing appropriate leaders to political offices and then helping them with decision making is crucial. Raising public awareness (including government officials and politicians), implementing conservative measures, considering various energy storage methods, identifying and utilizing renewable energy resources, implementing distributive generation, and beginning the process of converting to a hydrogen economy could happen more smoothly and more quickly if supported by national laws, policy, and financial incentives, such as tax breaks, and having subsidies for sustainable energy sources.

Change at this level also needs to acknowledge the interconnected relationship among the various areas of the United States. Hurricane Katrina can be used as an example. If distributive generation would have taken place on a wide scale, then increasing energy prices would not have affected so many people at the same time. Without the increasing prices of energy weighing heavily on families and communities, the devastating impact of Katrina could have been dealt with more appropriately and with less concern over individual energy needs and costs. Individuals and communities could have increased their financial support if not burdened by increasing energy costs.

A specific example of social change at this level can be seen in some of the policies we have made in the United States, such as offering tax incentives for purchasing hybrid vehicles. Unfortunately, our approach is not all inclusive. For such an example, countries like Sweden and Iceland offer more comprehensive approaches to swift social change concerning sustainable energy.

Iceland's national policies are much more sustainable because their initiative is to become the world's first hydrogen society. The creation of a hydrogen economy is official government policy, and their first success has been emission-free fuel cell buses. They also support a plan for nationwide hydrogen cars. They offer a perfect example of what can happen when support is widespread.

Change at the International Level

Although change at this level is not part of the discussion related to the six-step approach, a few important aspects of social change should be pointed out. First, when individuals, communities, and/or nations make social changes around energy, it impacts the international level. If local renewable resources are utilized, then more resources are available globally. Second, when renewable sources are used and pollution is decreased, international pollution is decreased. Finally, when social change happens on one level, it often impacts and spreads to other levels. If social change is made at the national level, then international impacts are sure to happen.

In conclusion, people often think social change means giving up individual rights. In fact, it means protecting and prolonging those rights. Generations to come have the right to survive with a quality of life similar to our own. People in other countries have the right to exist without suffering from our lifestyles. We can make some simple changes that would create a more sustainable society. We need to decide if we are going to give the next generations a gift or a problem. We need to decide if we are going to continue our global exploits or become part of the global community in a respectful and responsible manner when it comes to energy.

Appendix A

Various Estimates for the Amount of Petroleum Remaining

Perhaps no topic in the arena of petroleum supplies is more misunderstood or with more false information passed around than that of the amount of petroleum resources remaining. Over two thousand years ago the Greek Demosthenes stated, "Nothing is easier than self-deceit for what each man wishes, he also believes to be true." This certainly seems to be true of petroleitus's desire for an infinite amount of petroleum to feed our voracious appetites. This is simply not the case, however, and in this appendix, we will look at the evidence for the amount of reserves remaining. We will examine this evidence from three different perspectives. First, we will look at the Hubbert method and analyze the mathematics behind his projections and at Deffeye's simplification of Hubbert's method. Second, we will look at the U.S. Geological survey from the year 2000, along with the cornucopian view, and third, we will examine R. Oguz Capan's discussion on the amount of petroleum remaining.

Hubbert's Curve

In March of 1956, M. King Hubbert gave a presentation at the spring meeting of the Southern District Division of Production of the American Petroleum Institute in San Antonio, Texas. The presentation, entitled "Nuclear Energy and the Fossil Fuels," was an attempt to forecast the amount of the various types of energy sources left for worldwide usage. It was mostly disregarded at the time but since then has shown to be quite accurate in several of its predictions. With the data available to Hubbert at the time, he projected the United States would hit its peak production around 1970, and that is precisely what happened. He predicted the world would hit its peak production around the mid-nineties. In 1969 he published an updated prediction that was based on total world petroleum resources of 2.1 trillion barrels and peak production in the year 2000. The projections, off by a few years, were, however, amazingly accurate for the data available at the time. Some of the numbers used by Hubbert in his presentation have been updated since 1969 to more accurately refine his predictions.

What follows is a simplified version of Hubert's method. It is quite easy to get lost in his use of terminology, but his use of mathematics is not terribly difficult. First, we will look at Hubert's mathematics and then briefly look Kenneth Deffeye's interpretation of Hubert's method into a linear mathematics approach.

To understand Hubbert's method, let's first look at some basic math. If we are using some commodity at a certain rate, say ten oranges a day, over a five-day period, we will use five days times ten oranges per day, or fifty oranges. We can represent that by the use of a graph, where the vertical axis is the amount of oranges per day (the rate of orange use) and the horizontal axis is the total time we use the resources at that rate. We see that if we draw a horizontal line at ten oranges per day and a vertical line at five days the graph will map out an area enclosed by the two lines. This area is precisely equal to the total resource used, or fifty oranges (diagram 1). The process is very easy to see when we have a constant rate of consumption, e.g., ten oranges per day. The same relationship holds true, however, if we use the commodity at a varying rate. For example, the first day, seven oranges are consumed, the next day, ten oranges are consumed, the next day, twelve oranges are consumed, the next day, six, etc. If a varying amount of the commodity is used, the horizontal line will no longer be straight but will have a curved shape. Still, the area under the graph will be approximately the total amount of the commodity used (diagram 2). Using this concept, if we can find the shape of the curve for a given rate of commodity usage and we know how long the commodity will be used, we can then calculate the total amount of the commodity needed.

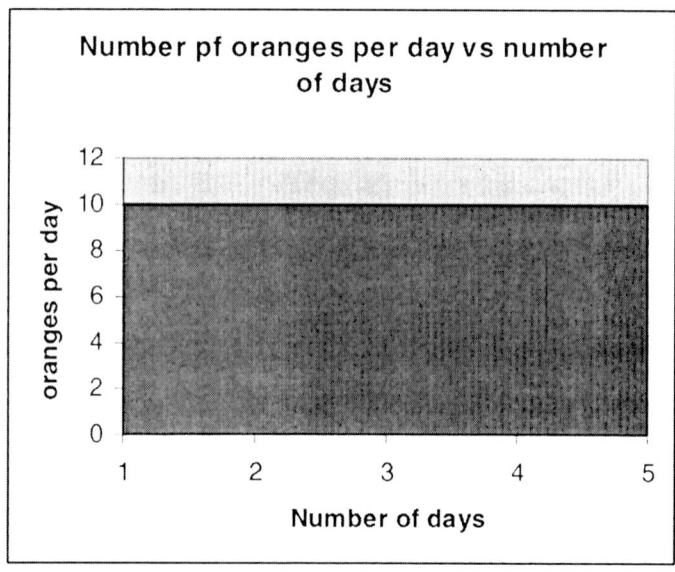

Diagram 1. Rate verses time of orange consumption. Area under graph is total consumption of the resource.

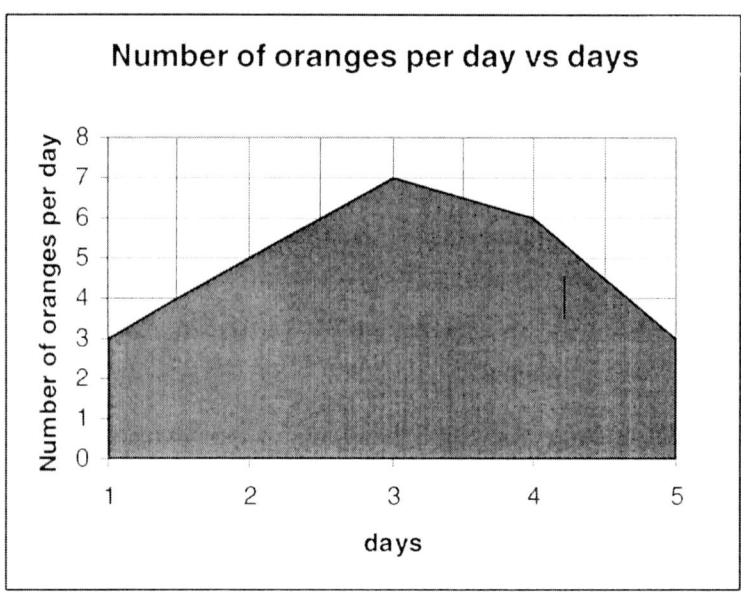

Diagram 2. The relationship still holds if we use a varying amount of the commodity. The area under the graph is still the total amount of the resource used.

Hubbert was faced with the reverse problem. He felt he had a good idea of the shape of the curve of petroleum usage (or any finite resource), and he had completed extensive calculations of the total petroleum available in the United States and in the world. Hubbert needed to fit those two pieces of data together to determine when we would hit peak oil production and when we would run out completely. In 1976 in a presentation at a World Wildlife Fund's conference entitled "Exponential Growth as a Transient Phenomenon in Human History," Hubbert more fully discussed the reasons and evidence for the shape of a graph describing the production and use of a finite non-renewable resource. He felt the resource would initially be produced in small amounts until efficient extraction and production methods were developed and more uses were found for the product. At that time, production would increase to exponential growth (this simply means that over a period of time, production rates would begin doubling) until either the cost of production begin to increase prohibitively or competing resources begin to infringe on the market. At that time, production would level off, begin to decrease, and gradually return to zero when production cost became completely prohibitive, the resource supply was exhausted, or a demand no longer existed. Hubbert tested this idea by examining production rates of petroleum reserves in certain areas that had already become exhausted. Virtually all followed, with minor deviations, the shape of the curve he had assumed for production of a finite non-renewable resource. The shape of this curve is shown in diagram 3.

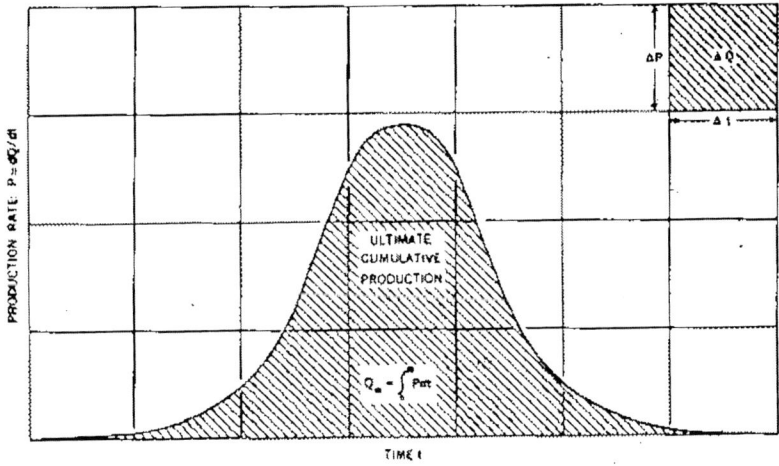

Diagram 3. The shape, over time, for the production and use of any finite non-renewable resource. The area under the curve represents the total amount of the resource available for use. Taken from Hubbert's paper presented in 1976 to the World Wildlife Fund.

Hubbert then fit the data he had accumulated on the consumption of petroleum to 1956 and the projected amount of petroleum remaining, both in the United States and in the world, and arrived at his curve. The U.S. production showed peak output in 1970. The world production showed a peak output in 1995 and is shown in diagram 4. Again, the total area under the graph is equal to the total petroleum available. Hubbert's last estimate in 1969 put the total world production at 2.1 trillion barrels and the peak production around the year 2000.

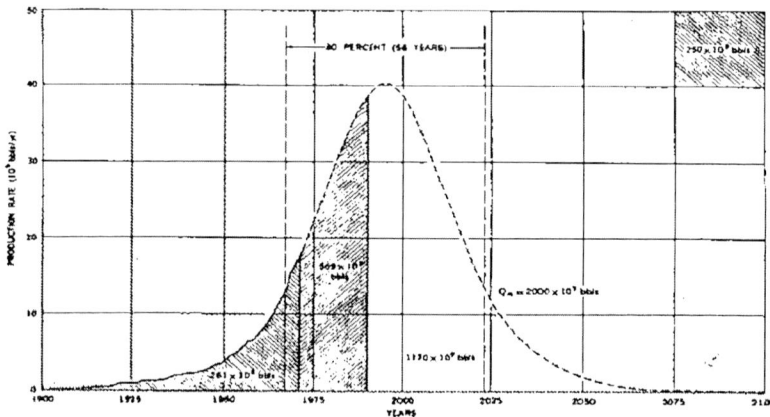

Diagram 4. Hubbert's initial curve for world oil production. Peak around 1995. Taken from Hubbert's presentation in 1976.

Hubbert's method is really quite straightforward and easy to understand. The difficulty appears to come in his use of terminology based in calculus. Quite often members of the non-mathematical public began to panic when they see an integral sign.

Deffeyes' Interpretation of Hubbert's Curve

In his excellent book, *Beyond Oil – The View from Hubbert's Peak*, Kenneth Deffeyes attempts to simplify Hubbert's mathematics by reducing it to a linear equation. Personally, the authors find Hubbert's method a bit clearer, but for many the straight line on a graph is much easier to understand than a curved graph. We will not reproduce Deffeyes' graphs here but according to his math the U.S. total petroleum production will max out at 228 billion barrels. This is in contrast to the U.S. geological survey of the year 2000, which gives an estimated U.S. total of 362 billion barrels. The USGS estimate will be discussed later. It looks as though Deffeyes is quite accurate in his estimate of total U.S. reserves. In reducing Hubbert's work to a straight line, that line appears to be far enough along that extrapolating it to the total reserves available is not much of a stretch. The US has already used up most of its oil.

The problem with Deffeyes' method when applying it to the world is that the world has only used up about one-half of the total reserves. Extrapolating his line through that distance may lead to grievous errors. Deffeyes does make this stretch and arrives at a total world production of around 2.013 trillion barrels and the peak production arriving around late 2005. Several events could change the slope of his line and the shape of Hubbert's curve. We will discuss the feasibility of one or more of these events happening in the next sections.

United States Geological Survey of 2000

In 2000 the USGS, after a study involving one hundred person years, gave an estimate of slightly over three trillion barrels of the world endowment of oil and no mention of a peak. This would about double the remaining world reserves. This looks quite promising, but even if the estimate were true, at the rate the world demand is increasing, it would only add approximately ten to fifteen years on to the time before we hit the problem of demand exceeding supply. One of the problems with the USGS study, however, is they relied on estimates given by the various petroleum companies and exporting countries. Since the publication of their results, at least one company (Shell) has stated they overestimated their reserves by a factor of 20 percent. Also for some unknown reason, in 1990 Saudi Arabia's officially reported oil reserves jumped by a factor of 49 percent and they have never reduced that number since 1990. Other countries in the Mideast, Kuwait, Iran, and Iraq, also suddenly increased their supplies around the same time, from 39 percent to well over 100 percent. It is as if they are using their oil but not diminishing their reserves. If we could apply that concept to all world reserves, we would never run out—in fact, we would end up with more than we started with. Also included in the USGS total is the yet undiscovered oil, which is basically a shot in the dark. Again, much of this data is dependent on accurate reporting from the various companies and countries and is highly suspect.

Cornicopian View of Energy Reserves

Basically, the cornicopians say not to worry, technology will bail us out. First, energy use will become more efficient. Look at what happened with the automobile, computers, mass transit, and communications. Second, cornicopians like Huber and Mills in their new book, *The Bottomless Well: The Twilight of Fuel, the Virtue of Waste, and Why We Will Never Run Out of*

Energy, feel that energy is not the issue but the ordering of energy. They somehow interpret the first and second laws of thermodynamics as saying we will never run out of energy as long as the universe is around. They agree with other cornicopians that technology will come along that allows us to develop the massive energy in ocean hydrates, Canadian tar sands, and western oil shale. They neglect to say that technology has been trying to accomplish this for forty years and have not really got anywhere. Also, they look at improvements in retrieval technology in removing oil from wells that we can reopen. As Deffeyes points out, there is simply not much evidence to support these views and also, they ignore the environmental impacts produced from the burning of fossil fuels.

R. Oguz Capan's View

R. Oguz Capan is an oil and gas production engineer and consultant. In a very insightful PowerPoint presentation placed on the web, entitled "What Is Going On in the Global Oil Sector," he shows that in spite of increased world demand, primarily by the U.S. and China, there is no increase in global oil tanker capacity, global refinery capacity, or global recoverable reserves. Even though, if the oil was there, large profits could be made in any of these areas of building oil tankers, building refineries, and exploring for more reserves, there is simply nothing being done. As Deffeyes states, "It is conspicuous by its absence." In a free-market economy, there is no reason for this lack of enterprise if the oil was there.

Put very simply, if the majors could make money searching for oil, they would find the oil, build more refineries, and build more tankers to supply those refineries. In 1999 the ten largest oil companies spent $6.5 billion on exploration and discovered $18 billion worth of new reserves. In 2002 they spent 7.2 billion on exploration and found $5.5 billion worth of reserves. This negative trend has widened in 2003 and 2004. With the increase in oil prices over the past few years, the negative gap would have narrowed but not enough to promote large-scale exploration. With the new technologies available for exploration, it is not so much a throw of the dice to find oil. Two new fields that are being touted by the press as major finds are the Sakhalin-5 field in eastern Siberia and the "giant" Kasagan in Kazakhstan. With most optimistic reserve estimates, the total petroleum available in both fields will supply the world demand for oil a maximum of seven months. If ANWAR in Alaska were tapped, it would only add another five months to the world supply. The globe has been searched, and there are not many new reserves available.

After examining the evidence for the past five years, the authors primarily agree with Hubbert and Deffeyes' description of our energy future. It looks right now that virtually every major oil-producing area in the world is going all out with no more reserves. Saudi Arabia claims to have reserves but when asked by President Bush to expand production to cover the losses from the 2005 summer hurricanes, they were unable to do so. As this is being written (January 2008), President Bush is again asking Saudi Arabia to increase oil production, and they are again cold to the idea. They blame it on their infrastructure not being able to handle any greater production, but they are not expanding that infrastructure. It now looks that we are at Hubbert's peak and that within the next few years, the demand should begin exceeding the supply for sweet crude oil unless the world immediately begins a massive policy of energy conservation. It is the only way we feel the immediate effects of an oil shortage could be mitigated. If an energy conservation policy were implemented nationwide and worldwide, we could flatten Hubbert's peak, which could allow us time to enter a sustainable future.

Appendix B

The Thermodynamics of Heat Engines
First Law and Efficiencies of Heat Engines

This appendix is certainly not required reading for understanding the material in this book. It is put here by Dr. Jacobs in the simple hope that those readers who wish a more complete understanding of the whole energy situation may also wish to have a better understanding of the two basic laws that govern all energy interactions.

My first experience with thermodynamics left me with an absolute fear of the subject. While a young university engineering student, I had a thermodynamics professor who left much to be desired in the art of teaching. He would write partial differential equations on the board with his right hand while talking to the board and erase the equations as he passed them with his left hand. I learned nothing. For years, if someone said the word "thermodynamics," I thought they were swearing at me.

Over the years, I finally worked up the courage to begin a process of self-study in that area. I then found it to be an elegant, logical, and not terribly difficult subject. We will try to present the material in this appendix in a fairly logical and historical manner that we hope will be easy to understand. Math will be used to help explain the topic but in a clear and simple manner.

Most of the early work on thermodynamics was accomplished before scientists knew that matter was made up of molecules, atoms, electrons, protons, and neutrons. Initially, scientists thought heat was a fluid that flowed from a hot body to a cooler one. This perception of matter, beyond the atomic, is called the macroscopic approach to nature. We will approach thermodynamics from that point of view and not even discuss atoms, molecules, electrons, protons, or neutrons. In our discussion of thermodynamics, we will simply look at various heat systems. A system is defined as whatever you are working on and can be a closed system that allows no mass to pass in or out of its borders or an open system, which allows an exchange of mass across its borders.

The first step in discussing a subject is to become familiar with the definitions and units used to define properties that are discussed in the subject, energy, work, heat, volumes, areas, and forces. As we move into unfamiliar areas, we can learn much by looking at the units we come across.

Science has no good definition of energy. It is a fairly abstract concept, and we really do not know what it is. Lower-division science books will define energy as the ability to do work. I have used that definition most of my teaching career but do, however, tell the students it is not a very good one. Can energy do other things besides work? If it cannot do work, is it still energy, and on and on. We do know the amount of energy contained in a system is a function of various properties of that system, e.g., position, temperature, speed, pressure, volume, and so forth. So although we do not really know what the energy of a system is, we know we can change the amount of energy in a system by changing those properties of that system. As we will see, changing the energy of a system by changing its properties allows us to find units for energy.

There is a much easier definition for work. Scientists have defined work as force acting through a distance. In mathematical form, we can write:

Work = Force times distance, or short form: **W = F x d**

If we apply force to an object that does not move, we have done no work. Also, if an object continues to move with no force applied, again, we are doing no work. If we push an object through one foot with one pound of force, it is call one foot pound of work or a ft-lb of work. If we push a force of one Newton (unit of force in the metric system) through as distance of one meter, it is called a Newton-meter, which is now called a *joule* of work.

Early definitions of heat were arrived at through the amount of heat needed to raise the temperature of water. A BTU is defined as the amount of heat needed to raise the temperature of one pound of water one degree F. One calorie is defined as the amount of heat needed to raise the temperature of one gram of water one degree Celsius. A kilocalorie (usually large **C**) is defined as the amount of heat needed to raise the temperature of one thousand grams (one liter) of water one degree Celsius. In other words, from the macroscopic point of view, we really do not know what heat is, but we do know it can raise the temperature of some system, so we have defined it in those terms.

Volume is simply the amount of space contained given by three dimensions, usually length, width, and height. The units will be length of those sides cubed, or l^3. Area is just length, and width and the units are length squared, or l^2.

Finally, we will look at temperature scales and examine the Celsius and Kelvin scales, which are the two primary scales used in science. Celsius takes the zero point as the freezing point of water and one hundred as the boiling point of water. The increments are simply those evenly divided from zero to one hundred. As scientists learned more about heat, they found there was still a large amount of heat in objects that had temperatures below $0°$ Celsius. They then developed a temperature scale that started from a point where no heat existed (I realize I said they would not be mentioned, but no heat simply means that atoms and molecules have no vibrational motion. At absolute zero, all atomic motion stops). It is called the Kelvin temperature scale and starts at absolute zero. Each increment is the same as on the Celsius scale, so it takes the same amount of heat energy to increase the temperature of water one degree Kelvin as it does one degree Celsius. Approximately $-273°$ Celsius is the same temperature as $0°$ Kelvin.

At the end of the eighteenth and the beginning of the nineteenth century, scientists began looking at heat and energy transfer. They found that if two masses of different temperatures were put in contact with each other, the heat lost by one mass was equal to the heat gained by the other mass. We have all experienced this by adding hot water to cold water and the temperature of the mixture arriving somewhere in between the two.

Also around that time, Benjamin Thompson, who became Count Rumford and then Lord Kelvin, was boring cannon and noticed the more work it took to bore the cannon, the hotter the cooling water became. This seemed to show a relationship between work and heat. Fifty years later, Joule, a Manchester brewer, designed an apparatus to determine this relationship. The design included two paddlewheels under water, enclosed in a well-insulated container, that were connected to two weights that when allowed to fall would turn the paddlewheels. He could then equate the change in temperature of the water with the work done on the falling weights. He found that 4,200 joules of work could raise the temperature of one liter of water $1°$ K. **Therefore, 4.3 kilojoules = 1 Calorie**. This is the mechanical equivalent of heat. From Joule's work, we can now describe the units of heat added to a system also in terms of joules.

Also from this early work, there began a sneaky suspicion that in any reaction involving the heat, work, and energy contained in, added to, or taken out of a system, there was no net loss or gain, simply a rearrangement of those three properties. From this suspicion, the first law of thermodynamics was formulated. This can be written as:

The change in internal energy of a system is equal to the work put into the system or the work done by the system and the heat added to or taken out of the system.

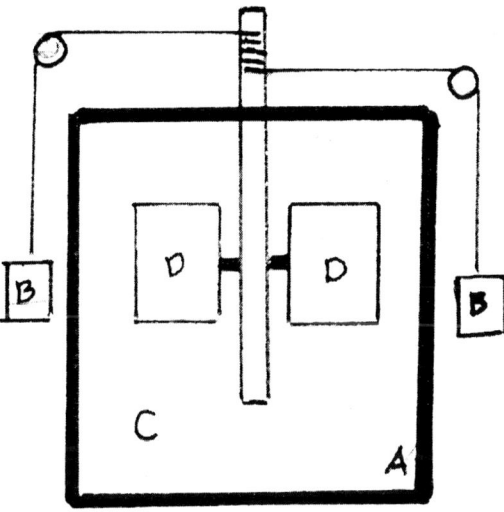

Figure B1. Diagram of Joule's apparatus. Water, C, is enclosed in a well-insulated container, A. The paddlewheels, D, are rotated as the weights, B, fall. As the weights fall, the water is agitated with an increase in temperature.

This statement can be written more concisely in a mathematical format as

$$\Delta U = Q - W$$

In this equation Δ means "change in." From now on whenever you see the Greek letter Δ, just say "change in." U is the internal energy of the system. Q is the heat added to or taken out of the system, and W is the work done by or on the system. The minus sign in front of the W is there simply because the founding fathers of thermodynamics, somewhat arbitrarily felt a heat engine should do work; therefore, work should be positive when going out of the heat engine. For this to be mathematically correct, they had to put the minus sign in front of work (W). Again, we do not know what energy is, but we do know it is related to heat and work and can be changed by the addition or subtraction of each and therefore must have the same units, which are joules.

We have not derived the First Law of Thermodynamics. The first law has no proof. It was simply stated in the middle of the 1800s and basically dared anyone to prove it wrong. In the last one hundred fifty years, although it has been modified somewhat by the discovery of new types of energy, no one has been able to disprove it, and it is now accepted as one of the cornerstones of modern science and engineering

With the first law, let us now examine a system that is used in many heat engines, that of a piston moving in a cylinder. In changing this system, we will have a force pushing against the piston rod, which decreases the volume in the cylinder. In doing this operation, we will look at the concepts of equilibrium, reversibility, and non-reversible. If the system is not changing and all of the properties, e.g., temperature, pressure, etc., are the same throughout the system is said to be in equilibrium. If any change takes place extremely slowly and in extremely small increments that allows equilibrium to be arrived at for each small increment, the change is said to be reversible. The changes can then be reversed with no net loss in heat or energy to the outside. If the system undergoes a rapid compression or expansion, heat is generated in the system or lost from the system and the pressure may not be uniform throughout. The rapid compression and expansion is a non-reversible process. We have all felt this heat when rapidly pumping up a bicycle tire.

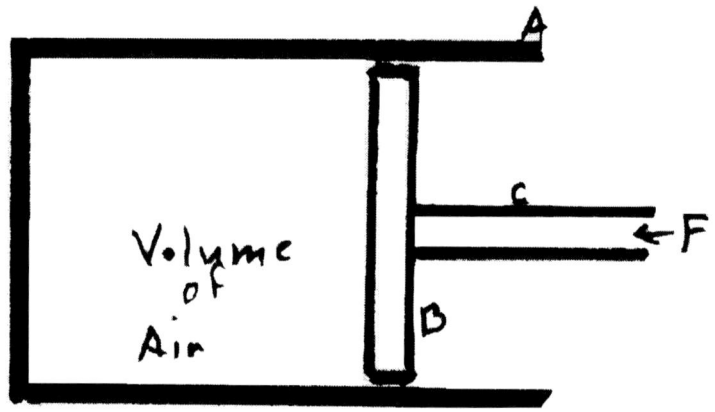

Figure B2. Diagram of piston B and cylinder A. Connecting rod C pushes the piston against the volume of air with a force, F.

This slow change (reversible process) in the piston and cylinder system is a purely theoretical situation and is impossible to achieve in nature. It would take an infinitely long time. It was, however, the genius of Carnot who first envisioned this theoretically slow process and stated in a piston and cylinder system, a reversible process had to be the most efficient compression and expansion cycle possible with no loss of energy.

When the piston is forced further into the cylinder, work is done on the system equal to force x distance. This distance pushed will produce a change in volume of the gas in the cylinder equal to d times the cross sectional area of the cylinder, A, or:

$\Delta V = d \times A$. If we solve this equation for d, we get: $d = \Delta V/A$.

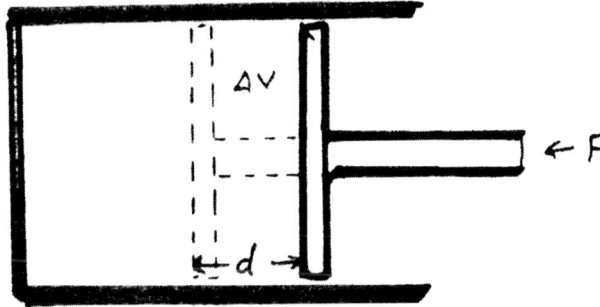

Figure B3. Force F pushes the piston against the trapped air and moves A distance D.

The change in volume is ΔV.

Therefore, work = force x distance, or: $w = f \times d$. Substituting the value for d, we get $W = f \times \Delta V/A$.

When the piston is pushed inward, we do work by increasing the pressure on the gas. Pressure is defined as force per unit area of the cylinder, or $P = f/A$. Solving this equation for f, we get: $f = \Delta P \times A$.

If we now substitute that value for f into the previous equation, we get: $W = (\Delta P \times A) \Delta V/A$

Here we see the A in the numerator will cancel with the A in the denominator, and our final equation is: $W = \Delta P \times \Delta V$.

The work put into the piston and cylinder system is therefore equal to the increase in pressure of the system times the decrease in volume of the system. We will now make a graph of the operation by plotting pressure on the vertical axis and volume on the horizontal axis. We start at some pressure P_1 and volume V_1 and end up at some pressure P_2 and volume V_2.

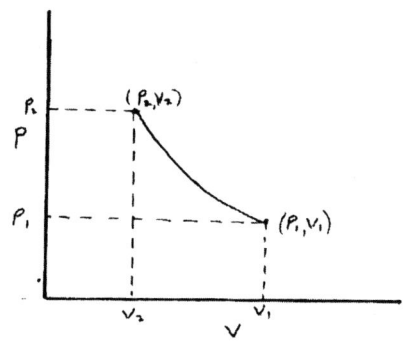

Figure B4. Pressure/volume graph of piston moving in cylinder with initial pressure and volume of $P_1 V_1$ and final pressure and volume of system $P_2 V_2$.

We are doing this as a reversible operation and therefore must do it in very small and slow increments as shown in the diagram below. We will notice that the area of each one of the increments is simply width times height or change in pressure times change in volume over each one of the small increments. As we have shown, this change in volume times change in pressure is simply the work done over that increment. If we could add up all the incremental areas on the graph, we would have the total work put into the system by the piston.

There is a mathematical process called integration that allows us to add areas under a graph. For our purpose here, we do not need to learn integration or calculus. We will simply use the Greek letter Σ to symbolize the adding of the areas under the graph. The total work done on the system during the compression of the cylinder therefore equals the total sum of the areas under the graph, or: $W = \Sigma \Delta P \Delta V$

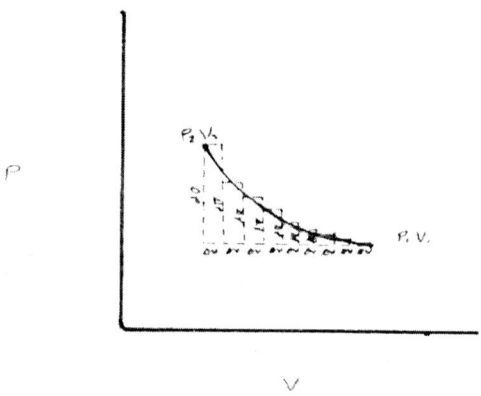

Figure B5. PV graph showing that small ΔP times ΔV rectangles add up to the total work done by the system.

Now that we have an understanding of the First Law of Thermodynamics, let us look at Carnot's work in terms of the maximum efficiency a heat engine can obtain. As was mentioned in chapter five, efficiency is defined as the ratio of the amount of work that an engine supplies to the amount of energy that is put into the system to supply that work.

Efficiency (in percent) = W/Q times 100

When Carnot first published his work in 1824, as we saw in chapter two, the steam engine was just starting to make its mark. Watt had greatly improved the efficiency of the machine, and Carnot wondered if there was any maximum to the efficiency a heat engine could obtain. He knew if a hot reservoir and a cool reservoir were thermally connected, they could arrive at a medium temperature without any work being done. If, however, a heat engine was placed between the hot and cold reservoirs, the heat engine could accomplish work.

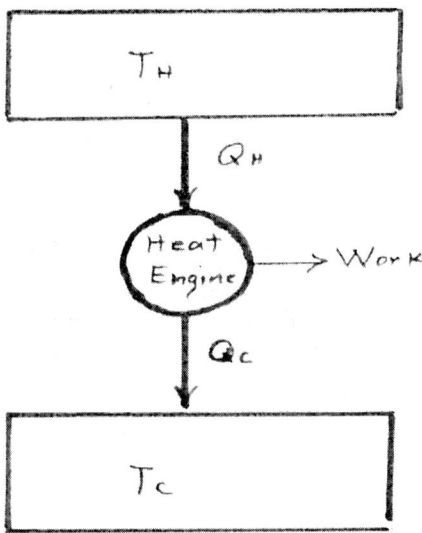

Figure B6. Diagram of a heat engine between hot and cold reservoirs of temperature T_H and T_c.

Carnot realized real steam engines leaked steam and heat energy and had large amounts of friction losses but in calculating the maximum possible efficiencies, he ignored these possible losses and determined the engine would pass through only reversible processes. He imagined a hot reservoir maintained at a constant temperature, T_h, and colder reservoir (usually the surrounding air temperature), T_c. The engine must operate continuously in a cycle where it will return to its initial state to repeat the process. During this cycle, it must take in the heat energy from the hot reservoir at the temperature, T_h, use this heat energy to accomplish work, discard any excess heat energy at the cooler reservoir, T_c, and finally return to its initial position to begin the cycle over. We will draw this cycle on a pressure volume graph and examine each aspect.

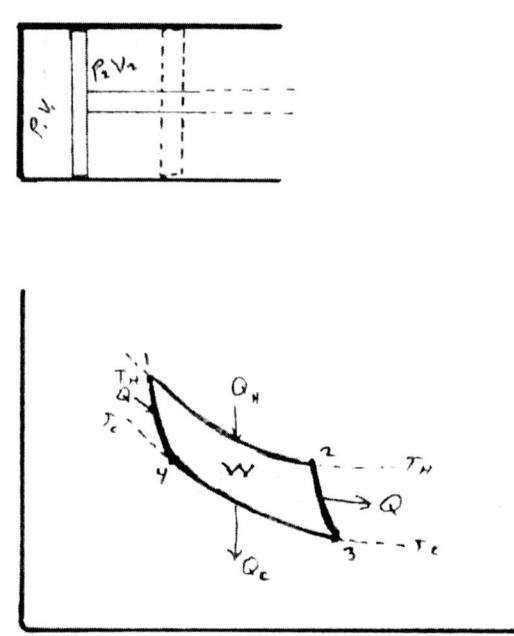

Figure B7. Diagram of piston in cylinder (above) slowly moving through the Carnot cycle on a PV diagram (below).

The cycle begins a point 1 with the piston at the inner limit of its stroke; therefore, pressure is greatest and volume is lowest. The piston is pushed outward while taking in heat energy Q_h from the hot reservoir at T_h to maintain a constant temperature. If the process is to be reversible, it must be done at a very slow rate at a constant temperature, and work will be done on the piston while the gas is expanding, moving the piston to point 2. At point 2, the gas is cooled to T_c (point 3) and heat energy Q_c is expelled to the colder reservoir. Work is then put into the system to slowly return the piston to its starting position while maintaining temperature T_c so as to be reversible. At point 4, heat energy is again added to bring up the temperature to T_h. The process is then repeated.

As we demonstrated earlier, the total work done by the system is the area enclosed under each movement of the piston. The work out is the area under the graph from 1 to 2, and the work into the system is the work under the graph from 3 to 4. The net work done by the system is then the difference in these two areas, or the area within the closed cycle. If we apply the first law to this, we see the change in the internal energy of the system (heat engine) is equal to the difference in the heat energy in and the heat energy out minus the work done by the system. Mathematically, this can be written as:

$$\Delta U = Q_h - Q_c - W$$

Since the heat engine operates in a closed cycle, the initial energy of the system must be the same as the final energy of the system, or ΔU must equal 0. Therefore:

$$\Delta U = 0$$
$$0 = Q_h - Q_c - W \text{ or}$$
$$W = Q_h - Q_c$$

The efficiency is defined as work out divided by energy in, or Eff. = W/Q_h

Substituting the value for W

$$\text{Efficiency} = (Q_h - Q_c)/Q_h \text{ or Eff} = 1 - (Q_c/Q_h)$$

We can show that as the temperature T_h becomes hotter, the heat energy Q_h transferred to the engine is greater. Likewise, as T_c becomes colder, the heat energy Q_c transferred from the engine to the cold reservoir also becomes greater. We assume (but have not proved) that the ratio Q_c/Q_h may be equal to the ratio T_c/T_h, which it is and could be shown under a more formal proof. If the two ratios are the same, we can then replace the heat energy ratio, Q_c/Q_h, with the temperature ratio, T_c/T_h, or:

$$\text{Eff} = 1 - (T_c/T_h), \text{ or we could rewrite as:}$$
$$\text{Eff} = T_h/T_h - T_c/T_h \text{ or:}$$
$$\text{Eff} = (T_h - T_c)/T_h$$

This means the maximum efficiency possible is the difference between the temperatures of the hot and cold reservoirs over the temperature of the hot reservoir of the heat engine; therefore, the hotter the hot reservoir or, even more important, the colder the cold reservoir, the more efficient to engine will be. The cold reservoirs are usually the temperatures of the ambient air or water used for cooling, and the temperatures of the hot reservoirs are restricted by the materials used in constructing the engine. Again, this is an idealized heat engine and normal engines do not reach nearly this efficiency. Let us briefly examine some types of heat engine and look at their efficiencies.

Following Carnot's work with the steam engine, modifications of his cycle were envisioned, which led to the development of several other heat engines. In the 1876 Nikolous Otto, based on work by Rochas, developed the Otto Cycle internal combustion engine. Later the Rankin cycle was developed, which led to the steam turbine heat engine. Also based on Carnot's work, the Stirling engine was developed, which used a different working fluid than the steam engine.

In an internal combustion engine the air gas mixture combusts at around $900°$ Kelvin (T_h) and the ambient air temperature is approximately $27°$ Celsius, or approximately $300°$ Kelvin (T_c). This leaves a theoretically maximum efficiency of the perfect internal combustion engine of around 60 percent. This is not possible to achieve and with the heat loss and non-reversible cycles, the maximum efficiency of an internal combustion engine is closer to 30 percent. If we do the same operation on a steam turbine power plant, we get a theoretical maximum efficiency of around 47 percent and an actual efficiency of also around 30 percent. The closer to peak output a heat engine runs, the T_h stays at a higher level and therefore, the higher the efficiency of the system. What is noticeable in these calculations is that it is not possible, even with a perfect engine, to get near 100 percent efficiency. There will always be waste heat given off in a heat engine. The first law does not address this necessary waste in a thermal reaction, so we can assume it is not a complete description of the properties that affect heat engines.

The Second Law of Thermodynamics
Entropy

When scientists are not sure of something, we always start drawing pictures. (We like pictures.) Usually, these pictures are in the form of graphs, and we try to learn what we can from various types of graphs. We have already discussed the variables (properties) that make up the processes that go into a heat engine cycle. They are heat energy (Q), work (W), internal energy (U), pressure (P), and volume (V). We have already used a pressure-volume graph to show the cycles of a heat engine and have found the work done by the heat engine is the area contain within the graph. Now let's look at other graphs using the other variable to see if we can find anything of value in them. We will again use a heat engine and put it through nearly reversible cycles and then plot our values for the variables we measured. We would first set up a table of the values we have found.

P	V	T	Q_{rev}	W_{rev}
P_1	V_1	T_1	0	0
-	-	-	-	-
-	-	-	-	-
-	-	-	-	-
-	-	-	-	-
P_2	V_2	T_2	Q_1	W_F

Figure B8. Graph of values in make-believe experiment in which we move a piston and cylinder through various paths on a pressure/temperature diagram.

Our first graph will be pressure verses temperature, and we will take different paths between beginning point 1 and ending point 2.

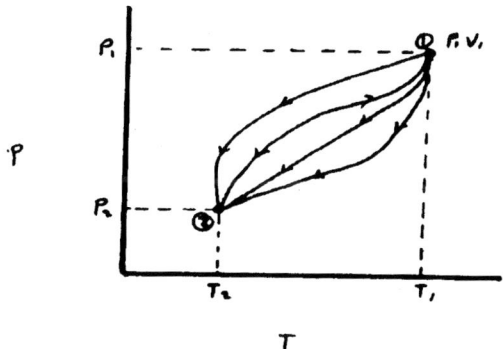

Figure B9. Make-believe pressure/temperature graph made from the make-believe data in the previous table.

From this graph, we notice that V_1 and V_2 are the same no matter what path we take. We realize the final volume is simply a function of the change in temperature and pressure and not function of how we got there. It doesn't really help us much because scientists had already figured this out in the later half of the eighteenth century (PV/T = Constant). We keep going and looking for other relationships. We soon figure out that although Q has different values for all the paths and W has different values for all of the paths, the difference between the two Q – W is the same for all of the paths. At first, this seems as though we have discovered something enormous, but we soon realize this is what the first law already tells us ($\Delta U = Q - W$).

Undaunted, we keep going and plot Q verses W, Q verses T, W verses T, W verses P, etc., etc., and we find nothing significant. Now we are getting desperate, and so we begin plotting the inverse of the variables, $1/T$ verses W, $1/W$ verses T, etc. Finally, at $1/P$ verses W, we find something. Again, the areas under these graphs are all the same, but soon we find it is just equal to the change in volume, or:

$$\Delta V = (1/\Delta P) \times W \text{ Or } W = \Delta V \times \Delta P$$

We already knew the work is equal to the change in pressure times the change in volume, so again, we keep looking. Finally, in the graph of $1/T$ verses Q, we find the area under the graph to be almost constant. Eureka! We have found a new property, and this time we have not seen it before. We know it has the units $(1/T) \times Q$ or Joules/°Kelvin, but what to call it? Since it is a property, we can call it xperty, but every new thing is called x, so let's call it Ntropy or entropy and give it symbol N. N is not acceptable to all scientists, but they can agree on s; therefore, this new property is called entropy and has the symbol s. We really don't know what it is, so we run a bunch more thermodynamic processes and measure the change in entropy in all of them. We look at heat pumps, refrigeration, steam engines, steam turbines, internal combustion engines, and your basic fire. It turns out the value for the change in s (Δs) is always positive and the further away we get from a perfectly reversible process, the higher the value for Δs becomes. Near reversibility Δs is almost zero. From our experiments, we can now state the Second Law of Thermodynamics:

In any thermodynamic process, the change in entropy is always equal to or greater than zero. Or mathematically:

$\Delta s = > 0$

Again, we are not sure what entropy is, but since it is always positive, we have a sneaking suspicion that it is related to the required inefficiencies in a heat engine. The law was first stated in 1865 and is now accepted as the second cornerstone of thermodynamics. Only those processes that satisfy the second law are possible in nature. If we can envision a process that does not satisfy the second law, it is not possible.

Although we are still not quite sure what this property is, over the years scientists have related it to amount of disorder in the universe, the higher the entropy the greater the disorder. Energy is required for order to be increased. Solar energy grows plants and people, which are in a high state of order. Mechanical energy can compress a spring or remove the dust from a room, which is again ordered. We use biological energy to straighten up a room. Solar and heat energy gave us the fossil fuels, which are an ordered form of matter. In any thermodynamic process, however, the entropy must be increased, which means an overall increase in disorder.

Over the years, this seemly insignificant property of matter has taken on very philosophical meanings. Some call it the arrow of time (the universe is always moving toward a state of greater disorder). If we go back in time, we must decrease entropy, which according to the second law is impossible. Whatever it is, entropy is a powerful tool in helping us understand the processes that occur in nature.

APPENDIX C

Is It Possible? Can We Go Sustainable?
The Amount of Renewable Energy Available

In discussing the path to sustainability with local groups, the question and major concern is usually, will the amount of renewable, sustainable energies be enough to supply all of our present and future energy needs? Our answer is "not only must we find the energy, but all evidence points to the answer 'yes'." The United States' share of the solar energy that strikes the earth's surface is forty thousand exojoules (10^{18} joules) per year. This is more than four hundred times the total energy consumed per year in the U.S. It would be very easy to just say, "See, there is enough energy, so let's do it." But let's break the major renewable technologies down and investigate conversion efficiencies and the amount of energy available. We will look at solar, wind, and biomass, which will probably be the big three in a new economy but certainly do not exhaust the renewable energies available at local levels that were mention in chapter five.

For our calculations in this appendix, let us take an average yearly national energy usage of fossil fuels to be about seventy to eighty exojoules. Of this approximately forty-one exojoules go toward the generation of electricity and of that, fossil fuels generate twenty-nine exojoules and of that, twenty-one exojoules are generated from coal. The first step in our conversion to a sustainable economy is to institute conservation measures, which are estimated to cut energy consumption approximately 30 percent. We will be a little conservative in all our calculations and assume it to be 20 percent. This will drop our national renewable energy needs to fifty-eight exojoules and that generated from coal to be about 16.8 exojoules. We will try to account for the fifty-eight total, and 16.8 exojoules from coal with conservative estimates of energy obtained from renewable, sustainable energy sources. Also, coal/steam generation of electricity is about 30 percent efficient when generated at remote generating stations, which brings the electric energy use in the U.S. that is generated from coal down to about five exojoules of energy.

As we are all aware, the amount of solar energy varies rather dramatically during the seasons and even quite often from day to day and during the day. It is, however, the most abundant renewable energy source we have and, if harnessed, could supply a large percentage of the energy needed. In chapter five, we discussed the various technologies that are involved in gathering solar energy. A tremendous amount of energy gain could be incorporated into passive and active solar systems, either retrofitted into existing building or designed into new buildings. If we look at a national average of six hundred watts of solar gain per square meter (about ten square feet) of collector, an average of ten hours per day of usable sunshine, and 50 percent of days overcast, in a flat plate collector (around 50 percent efficient) of five feet by ten ft (50 ft^2), we could collect 10^{10} joules (10^{-8} exojoules) of energy in a single dwelling. These are all very doable numbers. In some areas of the country, there would be greater gain and some areas less gain. Also, a fifty-square foot collector is quite small and on many buildings that size could be multiplied several times over. Multiply this by at least one hundred million (10^8) buildings we have in the United States, and we have begun to make a dent in our energy usage, about one exojoule per one hundred million, fifty square foot collectors.

Next let us look at areas of extremely high solar potential to see what energy we can get from photovoltaic cells placed in large arrays. We could also use concentrating collectors in some areas, but those usually need some type of cooling fluid, which is usually water and is in short supply in these regions. These desert regions have about two thousand five hundred kwh/m^2 of solar energy available annually. Both concentrating collectors and photovoltaic cells in the field have about 10 to 12 percent solar to electric conversion efficiency, so let us use the conservative 10 percent efficiency. For a one thousand-mega watt electric generation, it would take approximately 3.3 square mile of land area. This is not a terribly big area, a total of about 1.8 miles on a side and certainly doable. It could all be placed in one large area or over many smaller areas. If we multiply this times one hundred power stations in the desert or other high solar gain areas in the U.S., we get from three to four exojoules of energy. This alone would account for 60 to 80 percent of the energy needed to replace coal.

One very positive factor with photovoltaic is that there is no economy of scale. If we decide to install thousands of smaller systems, there would be no loss in efficiency. In fact, if these were placed near the end user, we would have less transmission losses that would make them even more efficient, less subject to large down time, and terrorism. This along with the solar placed on buildings would almost take care of the electric needs for the U.S.A. We are making a very large dent in overall fossil fuel energy usage.

Wind is probably the best near term source of electricity in the U.S. as far as technology available and cost per installed KW. It is now extremely competitive with coal-fired power plants and if environmental costs were included, it would be much less expensive. In 1991 the Pacific Northwest Laboratories conduct a study on the wind potential in the U.S.A. and found that with current efficiencies, wind energy conversion system could generate eleven thousand billion KWh of electricity (almost 40 exojoules). That is about eight times the amount of electricity generated by coal. In 2006 the U.S. only generated about 1 percent of its electrical energy with wind. The same study showed that five Midwestern states (ND, TX, KS, SD, and MT) could generate 5,570 billion kwh of electricity (about twenty exojoules or four times the amount from coal-fired electric power plants). This may turn out to be extremely beneficial to Midwest farmers. It looks as though climate change that is being brought on by the burning of fossil fuels will have drastic effects on the middle United State and will bring extremely dry conditions, which could possibly ruin the farm economy in those areas. The placement of wind generators in these states could be a major boon to the area.

The primary hindrance to the placement of the large wind farms is the lack of infrastructure needed to transmit the energy out of these areas. At this time, power companies are unwilling to install large transmission lines because of lack of national commitment toward sustainability. Besides the large wind farms, wind turbines of all sizes could be placed on rural residences and outside of towns to help furnish power in those areas. It appears the primary hindrance to these installations has been the lack of ability to tie into existing lines and the lack of economic backing of many state governments with tax credits or breaks to promote sustainability. All of these hindrances are certainly easy to overcome if we have the will to do so. Right now the powers that be in the coal industry have quite a lot of political power and are trying desperately to hold on to it. It does appear that in some states the massive threat of climate change is beginning to wake up the legislatures who are passing laws and regulations that promote sustainable energy.

Again, it is easy to say, "Look, we can get all that energy from wind, but how many wind systems is that and how will they be distributed?" The modern wind energy conversion system is eighty meters in diameter, placed on an eighty-meter tower and rated at 1.8 megawatt maximum output. The average output as a percent of peak load is called capacity factor and for most wind systems is in the range of 25 percent to 40 percent. If we split the difference and take a capacity factor of 32.5 percent, the system would produce an average output of .57 megawatts. That is approximately 1.8×10^{-5} exojoules annually; therefore, to get 1.8 exojoules, we would need one hundred thousand 1.8-megawatt wind systems scattered across the five state region.

If we start now, this would certainly make a major dent in our national electric demand and is certainly doable. If this was combined with thousands more wind systems of this size and smaller on farms and near towns, we could make a much larger dent. As more wind generators go on line, some should be dedicated to the production of hydrogen to supply energy when the wind systems are not producing enough power to fill the demand.

Biomass

By the term "biomass," we mean all living plant matter as well as all animal and organic waste, including sewage, dung, wood, and plant, and animal waste. Plants typically have conversion efficiencies from solar to chemical of about 1 to 2 percent. This seems extremely small until one realizes the tremendous amount of solar energy incident upon the U.S. per year.

Over the centuries, humans have relied on biomass as their primary source of energy and as human population has increased, a large number of world habitats have been destroyed as a result of over-harvesting. Because of this potential problem when using biomass, scientists have attempted to develop a formula for determining if these resources can be used in a sustainable manner. The simplest formula only takes into consideration the rate of usage to benefit the local population verses the rate of regenerating the resource. More elaborate formulas take into consideration the rate of production of dry biomass per unit land area, total area available for growing biomass, the unit of energy content of the dry biomass, the efficiency of conversion of the dry biomass into the usable form of energy desired, the population density of the area to be serviced, area of region to be serviced, the average per capita energy consumption rate, the fraction of the total energy consumed in which the biomass will be included, and the fraction of that total energy consumed that will be the biomass resource. No matter how complicated the equation becomes, it is very obvious there is a surplus of biomass, if harvested in a sustainable manner, that can make a large dent in our total energy use. In a study completed in 1991 and printed in 1993 in *Biomass Energy Supply Prospects* in "Renewable Energy: Sources for Fuels and Electricity," the authors estimate that if a policy pursuing a sustainable economy was implemented, by 2050, the North American continent could generate 40.7 exojoules of sustainable energy. The study included crops (1.7 exojoules), existing forests (3.8), animal dung (.4), biomass energy crops (34.8), and residuals (5.9). The residuals include forest, mill, and agricultural residues, urban wood wastes, municipal sludges/biosolids, and food residuals.

Most of this energy should be grown and used locally because of the low-energy density of most biomasses. Biomasses can be used for multiple types of energy production, including direct combustion for heating or electrical generation, gasification, fermentation into alcohol, and anaerobic digestion into methane (natural gas). Biomass can also be mixed with coal in fueling electrical generating plants and lower the sulfur and nitrogen content of the combustible material.

Summary

We have tried to show in this appendix that moving to a sustainable economy is certainly within the possibilities of current technologies. There is no new invention that has to come along, no dilithium crystals that have to be developed; we simply need the national desire to do so. The three sources discussed here will play a major role in that economy but certainly do not exhaust the possibilities of renewable energy sources that are available and discussed in chapter five.

Many of the powers that be (fossil fuel proponents) are opposed to this move and refuse to see the problems that are associated with the present fossil fuel economy. They give reasons for their views such as, it cannot be done on the massive scale that we use energy, there is no evidence for human-fostered global warming, and we must not affect the U.S. economy. Under closer scrutiny, none of these hold up. There is simply no dispute among scientists who study global warming that it is fact. As we have tried to show in Appendix C, it is very possible to move toward and, over the next thirty to forty years, arrive at a sustainable energy economy. Finally moving toward that economy could be one of best things that has ever happened to the U.S. economy. Without the payments for foreign oil, our national balance of payments would greatly improve. At a time when well-paying jobs are hard to find and unemployment is growing, massive new higher-paying jobs would be available in the gathering and processing of renewable energy. Even now in the biofuel and wind energy industries, there is major emphasis on finding and training new workers. This is just the tip of the iceberg of what could and should happen.

NOTES

Chapter 1

Page 4: There are many discussions in books, papers, and presentations on the amount of petroleum remaining. Kenneth Deffeyes in *Beyond Oil - The View from Hubbert's Peak* gives perhaps the most comprehensive description of M. King Hubbert's peak oil concept. The USGS offers a much brighter picture on the amount of oil remaining in their 2000 World Assessment Summaries, which can be found on the USGS website. The various pictures of the amount of petroleum remaining are discussed more fully in Appendix A.

Page 5: Amount of oil remaining in Alaskan wilderness and coastal California is discussed in the Hearing before the Committee on Energy and Natural Resources, United States Senate, July 18, 1995, page 62.

Page 5: China and India entering the world petroleum market is discussed by Dilip Hiro in *Blood of the Earth, Battle for the Worlds Vanishing Oil Resources,* Avalon Books, 2007, pages 183 to 211.

Page 6: 60 percent to 70 percent of U.S. oil imported. Senate Hearings, 1995 page 2.

Page 6: Wind energy is fastest-growing source of energy. Vital signs: World Watch Institute, 2005.

Chapter 2

Page 8: Amount of coal remaining. The Department of Energy on its fossil.energy.gov website claims there is from two hundred to three hundred years of coal energy remaining. Barbara Freese, in *Coal: A human History,* Penguin Books, 2003, states the amount of coal remaining is grossly overestimated and with increased usage there would be between fifty to one hundred years' supply.

Page 8, 9, 10: Barbara Freese discusses the beginnings of the coal revolution in England in chapter tow of *Coal: A Human History*.

Page 10, 11: Hero's treatise on Pneumatics, Translated by J.G. Geenwood, 1851, can be found on the web.

Page 11: Barbara Freese. *Coal: A human History*, pages 45 to 56.

Page 11, 12, 13, 14: The evolution of the steam engine from Huygens to Trevithick can be found on the web at *www.geocities.com*.

Page 14: Barbara Freese. *Coal: A human History*, pages 45 to 56.

Page 14, 15: Sadi Carnot, "Reflections on the Motive Power of Heat and Machines Fitted to Develop This Power," can be found on the Web at *www.history.rochester.edu/steam/carnot/1943/*.

Page 15: Edward Parson's article "The Steam Turbine" can be found on the web at *www.history.rochester.edu/steam/parsons /*.

Page 15, 16: There are many articles and books, both on the web and in the library on combustion of coal. One web source put out by the government is "Cleaning Up Coal," which can be found at *fossil.energy.gov*. An excellent source for the negative effects of coal-fired power plants is "Beyond Mercury: How the Fine Print of the Bush Administration Plan Means More Arsenic, Dioxin, Lead and Other Toxic Air Pollution" which can be obtained from the National Environmental Trust Fund.

Chapter 3

Page 17, 18: West Virginia Geological and Economic Survey gives the history of the early discovery of oil on their website. Also Daniel Yergin in *The Prize: The Epic Quest for Oil, Money, and Power*, Simon and Schuster, New York, 1991. pages 19 to 34 and Bill Gerber and Kenneth Anderson in *Modern Petroleum's Basic Primer of the Industry, 3rd Edition*.

Pages 19, 20, 21: Yergin, *The Prize*, pages 34 to 113.

Pages 21, 22: Any thermodynamics book.

Page 24: Yergin, *The Prize*, pages 248 to 252.

Pages 25, 26 : Reynolds, Clark G. *The Carrier War (Epic of Flight)*, U.S.A.: Time Life Education, 1982.

Page 26: Yergin, *The Prize*, page 326.

Pages 26, 27: Yergin, *The Prize*, pages 326 to 350.

Pages 28, 29, 30, 32, 33: Hughs, Barry; Rycroft, Robert; Sylvan Donald; Trout, Thomas, Harf, James. *Energy in the Golbal Arena*, Duke University Press, Durham, 1985, pages xiii to xvi; Yergin, *The Prize*, pages 541 to 664.

Page 34: William Shawcross in *The Shah's Last Ride*, Simon and Shuster, New York, 1988, pages 25 to 37.

Page 37: Leigh Glover in *Driving Under the Influence: The Nature of Selling Sports Utility Vehicles*, Bulletin of Science Technology and Society, October 2000, page 360.

Pages 37, 38: Vital Signs 2001, World Watch Institute, W.W. Norton, New York, 2001, pages 40,41.

There are many different organizations that compile data related to world petroleum resources. Some are based on very unrealistic estimates and some very conservative estimates. Various estimates are discussed more fully in Appendix A and in Kenneth Deffeyes' book *Beyond Oil: The View from Hubbert's Peak*, Hill and Wang, New York. 2005. Those discussed in chapter three are based on the data found in the following sources.

Page 39: World Energy Council of the United Nations Statistical Division, found on their web page.

Page 39: World Assessment Summaries, USGS world Assessment Team, USGS web page.

Page 39: Campbell, C.J. and Laherrce J.H. *The End of Cheap Oil*, Scientific American, March 1988, pages 78 to 83.

Page 40: Hearing before the Committee on Energy and Natural Resources, United States Senate, July 18, 1995.

Page 39: R. Oguz Capan in "What Is Going on in Global Oil Sector...?!" at *oguz.capan@solar-ltd.com*, December 2004.

Chapter 4

Pages 48-55: The development of the natural gas and accompanied legislation is discussed on the Natural Gas Association's website, in Herbert, John, Clean Cheap Heat: Development of Residential Market for Natural Gas in the U.S., Praeger Publishers, 1992.

Pages 50-55: Chambers, Ann, *Natural Gas*, Pen Well Publishing, 1999, discusses the legislation that affected the natural gas industry and the validity of the industry's claims as to how that legislation affected their industry.

Page 55: Defeyes, *Beyond Oil: The View from Hubbert's Peak*, Hill and Wang, 2005, Hiro, *Blood of the Earth*, Avalon Publishing, 2007, and Tertzakian, *A Thousand Barrels a Second*, McGraw-Hill Books, 2007, all agree on the U.S. and world supply of natural gas. We are peaking on the North American continent, with perhaps a twenty-year supply before peaking on the world market.

Pages 57-69: Jacobs, Dennis, *The Search for Science*, Ron Jon Press, 2000, El-Hawary, Mohamed E, *Electrical Energy Systems*, CRC Press, 2000, and Munson, Richard, *The Power Makers*, Rodale Press, 1985, all discuss the science and technologies needed for the development of the electrical industry.

Pages 60-64 Borbely, Ann-Marie and Kreider, Jan F., Editors, *Distributes Generation: The Power Paradigm for the New Millennium*, CRC Press, 2006, some articles, e.g., Wolfson Morey, "The Regulatory Environment," discuss the legislation that could lead to new electrical generation paradigms in the future. Other articles discuss the new technologies available for that shift.

Pages 64 to70: Munson Richard, *The Power Makers*, Rodale Press, 1985, pages 112 to117. Discusses the utilities and governmental development of atomic reactors and the accidents at Three-Mile Island.

Pages 67 to 68. There are a number of books and websites dedicated to the Chernobyl accident. The most impartial seems to be the *Information, Communication, and Networking Platform* websites, found at www.chernobyl.info.

Chapter 5

Hinrichs, Roger A. *Energy*, U.S. Saunders College Printing, 1992.

Huges, Barry; Rycroft, Robert; Sylvan, Donald; Trout, Thomas; Harf, James. *Energy in the Global Arena*, Durham: Duke University Press, 1985.

Borely, Anne-Marie; Kreider, Jan F., Editors. *Distributed Generation— The Power Paradigm for the New Millennium*, Florida, CRC Press, 2001.

Heinrichs, Robert A.; Kleinbach, Merlin. *Energy: Its Use and the Environment*, 3rd Ed. U.S.A.: Brooks/Cole, 2002.

El-Hawary, Mohamed E. *Electrical Energy Systems*, Florida: CRC Press, 2000.

Majumdar, S.K.; Miller, E.W.; Panah, A.J. Editors. *Renewable Energy: Trends and Prospects*, U.S.A., Pennsylvania Academy of Science, 2002.

Tester, Drake; Driscoll, Golay; Peters. *Sustainable Energy: Choosing Among Options*, MIT Press, 2005.

Johansson, Kelly; Reddy, Williams, Editors; Burnham, Laurie, Executive Editor. *Renewable Energy: Sources for Fuels and Electricity*, Island Press, 1993.

Chapter 6

Pages 89 to 91. Heinrichs, Robert A.; Kleinbach, Merlin. *Energy*, Saunders College Printing, 1992, pages 361 to 396.

Pages 89 to 91 Krauser, Jack J.; Ristenin, Robert A. *Energy and Problems of a Technical Society*, New York, 1988. Pages 367 to 425.

Heinrichs, Robert A., Kleinbach, Merlin. *Energy: Its Use and the Environment*, 3rd Ed. U.S.A.: Brooks/Cole, 2002.

Schrobert, Harold H. *Energy and Society*, New York: Taylor and Francis, 2002, pages 427 to 501.

Page 92 to 93, *American Academy of Pediatrics* (AAP), July 2001 Report.

Pages 92 to 93, Watras, Carl; Huckabee, John, Editors. *Mercury Pollution: Integration and Synthesis*, Ann Arbor, MI: Lewis Publishers, 1994.

Fascianna, Guy S. *Are Your Dental Fillings Poisoning You?* New Canaan, Connecticut: Keats Publishing, 1986.

Pages 92 to 93, Sigel, Astrid, and Helmut, Editors. *Metal Ions in Biological Systems, Volume 34*, New York, Marcel Dekker, Inc. 1997.

There is a tremendous amount of very good research on the web and in books and articles dealing with global warming. The following articles discuss global warming and its effect on the North Atlantic Current.

Weaver, Andrew J.; Hillair-Marcel, Claude. "Global Warming and the Next Ice Age." *Science*, Vol. 304, April 16, 2004.

Hakkinen, Sirpa; Rhines, Peter B. "Decline of Subpolar North Atlantic Circulation during the 1990s," *Science* Vol. 304, April 23, 2004.

Kerr Richard A. "A Slowing Cog in the North Atlantic Ocean's Climate Machine," *Science* Vol. 304, April 16, 2004.

Chapter 7

Pages 97 to 98, Casten, Thomas R.; Dowens, Brennen. Critical Thinking about Energy: The Case for Decentralized Generation of Electricity, Skeptical Inquirer, January/February 2005.

Pages 98 to 99, Editorial in *Scientific American*, February 2004.

Pages 98 to 99 Pimental, David, et al. *Renewable Energy: Current and Potential Issues, Biosciences*, Vol. 52, No. 12, December 2002, pages 99 to 100.

Pages 99 to 101, Schrobert, Harold H. *Energy and Society*, New York: Taylor and Francis, 2002.

Pages 99 to 102, Johansson, Kelly; Reddy, Williams, Editors. Burnham, Laurie, Executive Editor. *Renewable Energy: Sources for Fuels and Electricity*, Island Press, 1993.

Pages 101 to 102, Borely, Anne-Marie; Kreider, Jan F., Editors *Distributed Generation: The Power Paradigm for the New Millennium*, Florida, CRC Press, 2001.

Pages 102 to 103, Rifkin, Jeremy. *The Hydrogen Economy: The Next Great Economic Revolution*, U.S.A.: Penguin Putnam, Inc., 2002.

Chapter 8

Page 105, Page 111. Harper, Charles L, and Kevin T. Leicht (2002). *Exploring Social Change: America and the World* (4th Edition). Prentice Hall, Upper Saddle River.

Pages 109 to 110. Clean Energy Resource Teams (2005). "Helping Minnesota Communities Determine Their Energy Future." *www.cleanenergyresourceteams.org.*

Page 111. The Minnesota Project, University of Minnesota's Regional Sustainable Development Partnerships, Minnesota Department of Commerce (2003). *Designing a Clean Energy Future: A Resource Manual Developed for the Clean Energy Resource Teams.*

Pages 11 to 112. The Official Gateway to Sweden (2005). "Sweden's

Renewable Energy Resources." *www.sweden.se/templates/cs/Article____8796.aspx.*

Pages 111 to 112, Hirsch, Tim (2001). "Iceland Launches Energy Revolution." BBC News. *http://news.bbc.co.uk/1/hi/sci/tech/1727312.stm.*

Appendix A

Most of the information on M. King Hubbert's methods are taken from his presentation at the American Petroleum Institute's annual meeting in San Antonio, Texas, in 1956, entitled "Nuclear Energy and Fossil Fuels." The presentation can be found on the web.

Deffeyes Kenneth. *Beyond Oil: The View from Hubbert's Peak*, Hill and Wang, New York. 2005.

Capan R. Ogguz. "What Is Going On in the Global Oil Secto." Presentation can be found at solar: *ltd.com/presentations/english.pps.*

Appendix B

Most thermodynamics books will give derivations of the First and Second laws of Thermodynamics. The format followed in this appendix was similar to the one followed in Van Ness, H.C., *Understanding Thermodynamics*, McGraw-Hill, 1969.

Appendix C

Hinrichs, Roger A. *Energy*, U.S. Saunders College Printing, 1992.

Huges, Barry; Rycroft, Robert; Sylvan, Donald; Trout, Thomas; Harf, James. *Energy in the Global Arena*, Durham: Duke University Press, 1985.

Heinrichs, Robert A.; Kleinbach, Merlin. *Energy: Its Use and the Environment*, 3rd Ed. U.S.A.: Brooks/Cole, 2002.

Johansson, Kelly; Reddy, Williams, Editors. Burnham, Laurie, Executive Editor. *Renewable Energy: Sources for Fuels and Electricity*, Island Press, 1993.

Schrobert, Harold H. *Energy and Society*, New York: Taylor and Francis, 2002.

Majumdar, S.K.; Miller, E.W.; Panah, A.I., Editors. *Renewable Energy: Trends and Prospects*, Pennsylvania Academy of Sciences, 2002.